Fluid Power
Educational
Series

Cartridge Valves

Joji Parambath

Cartridge Valves

Copyright © 2026 Joji Parambath

All rights reserved

No part of this book may be reproduced or transmitted in any form or by any means, electronic or mechanical, including photocopying, recording, or using any information storage and retrieval system, without the written permission of the publisher.

ISBN: 9798320447247

https://jojibooks.com

First Edition 2020
Second Edition 2024
Revised Edition 2026

Disclaimer of Liability

The contents of this book have been checked for accuracy. However, we cannot guarantee full agreement since deviations cannot be entirely precluded. Only qualified personnel should be allowed to install and work on hydraulic equipment. Qualified persons are those authorized to commission, ground, and tag circuits, equipment, and systems in accordance with established safety practices and standards.

Dedicated

To my wonderful daughter-in-law, Hetal, whose grace and love continually inspire us. Thank you for being such a positive influence in our lives.

Table of Contents

Chapter	Description	Page No
	Preface	vii
1	Introduction to Cartridge Valves	1
2	Constructional Features and Circuits of Single-function Cartridge Valves	5
3	Constructional Features of Multi-function Cartridge Valves and Circuits for Check Function	15
4	Control Covers and Circuits for Directional Controls	19
5	Control Covers and Circuits for Pressure Controls	32
6	Control Covers and Circuits for Flow Controls	43
7	3-way and 4-way Spool-type Cartridge Valves	45
8	Actively Controllable Cartridge Valves and Circuits	48
9	Proportional Cartridge Valves and Circuits	55
10	Constructional Features of Integrated Manifolds	63
11	Typical Characteristics and Specifications of Cartridge Valves	66
12	Advantages of Cartridge Valve Systems	70
13	Applications of Cartridge Valve Systems	72
14	Installation and Maintenance of Cartridge Valve Systems	74
15	Objective Type Questions	76
16	Review Questions	77
Appendix 1	Symbols of Cartridge Inserts	81
Appendix 2	Symbols of Cartridge Covers	82
Appendix 3	Mounting Configurations of 4-port Hydraulic Directional Control Valves	84
17	References	91

PREFACE

With the introduction of cartridge valves in the 1950s, an essential innovation in hydraulic valve design began. The basic cartridge valve consists of an insert assembly that slips into a cavity machined into a manifold. Initially, the cartridge valve was intended to perform a single function; therefore, a cavity was designed to encompass the valve. Later, the cartridge valve technology grew to include multifunction and integrated circuit features, with many cartridge valves incorporated in a single manifold block. In recent years, cartridge valve technology has seen many improvements to reduce leakage, simplify design and construction, and increase reliability, efficiency, and cost-effectiveness.

This book provides an in-depth understanding of multi-function cartridge valves, including their concepts, configurations, and circuits for check, directional, flow, and pressure control functions. It also covers the active logic valves and proportional cartridge valves, as well as the constructional features of integrated manifolds. Additionally, it provides detailed information on the characteristics, specifications, advantages, applications, and maintenance of cartridge valves. The book is organized in an easy-to-understand manner, with many circuits given in multiple positions for quick comprehension. It is a great resource for anyone looking to learn more about cartridge valves.

The same author gives many other fluid power topics in other textbooks under the fluid power educational series. A list of all the textbooks is given at the end of the book. Also, please see the details at https://jojibooks.com.

Enjoy reading the book.
Your feedback is most welcome.

JOJI Parambath

Second Edition ...
The second edition of the book has been improved with practical insights to ensure it is up-to-date and relevant. Existing chapters have been restructured, and many chapters have been added, including active logic and proportional cartridge valves.

JOJI Parambath

Chapter 1 | Introduction to Cartridge Valves

Conventional hydraulic valves typically use spool-type moving elements within their valve bodies. They have threaded ports for making pipe or tubing connections. The conventional connection arrangement of a hydraulic system with plumbed components contains many interconnecting lines among these components. This cumbersome network of interconnecting lines increases component costs and installation costs, tends to be bulky, and promotes leakages within the system.

Further, the conventional valves' ports open and close simultaneously. This switching action promotes the development of shock pressures within the system.

Hydraulic systems that use large spool-type valves have lower system response than other modern systems.

Over the last few decades, many line-mounted hydraulic components have been replaced by integrated circuits, with poppet-type cartridge valves inserted into compact manifold blocks.

Cartridge valves that offer an alternative to conventional sliding spool valves hold a unique position in today's fluid power industry.

Cartridge valves have become integral to many fluid power systems due to their small size, fast switching speeds, and cost-effective design.

Cartridge valves are flexible, compact, and have fewer leak points.

The cartridge design approach for a hydraulic system offers many advantages, such as reduced weight, external fluid connections, lower space requirements, reduced leakage, lower cost, and higher efficiency compared to a system with individually plumbed components.

You may note that an advanced hydraulic system may incorporate a combination of the manifold-mounted spool valves and the cartridge valves.

This chapter systematically presents the fine details of the cartridge valves.

Fundamental Concepts of Cartridge Valves

A conventional two-port check (non-return) valve with an internal poppet permits fluid to flow readily in one direction but prevents it entirely in reverse. However, this valve can be redesigned to overcome its usual blocking action, allowing fluid to flow through it in both directions. This idea is the basis of the control concept in cartridge valves.

By definition, a cartridge valve's insert or logic element consists of only the internal moving element and can be considered an element without an integral housing. In addition to this insert, a cartridge valve system consists of a mounting block (manifold) with appropriate flow passages and a control cover.

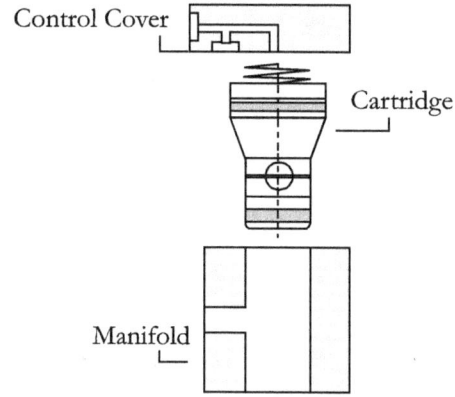

Figure 1.1 | The exploded view of the basic cartridge valve

An exploded view of the primary cartridge valve system is shown in Figure 1.1. The cartridge is inserted (slipped or threaded) into the manifold's standardized cavity. The control cover is placed over the insert and bolted to the manifold to retain the insert. Many types of standard covers are available with various control functions. The required control function can be achieved by combining the cartridge with an appropriate control cover. Remember, cartridge valves are designed to meet specific international standards.

Cartridge valves for directional and check functions are essentially hydraulically piloted check valves. They can also be designed for pressure and flow control functions. Each valve has only one control area in its spring chamber. However, there are logic valves, named active logic valves, each with a differential spool with two control areas.

Evolution of Cartridge Valves

Cartridge valve technology, originally conceived as a single-function design, has expanded to include multiple functions in a single cartridge valve. Such a cartridge valve is usually categorized as a multifunction cartridge valve.

Further advancement in cartridge valve technology has led to the development of an integrated manifold with multiple cartridge valves.

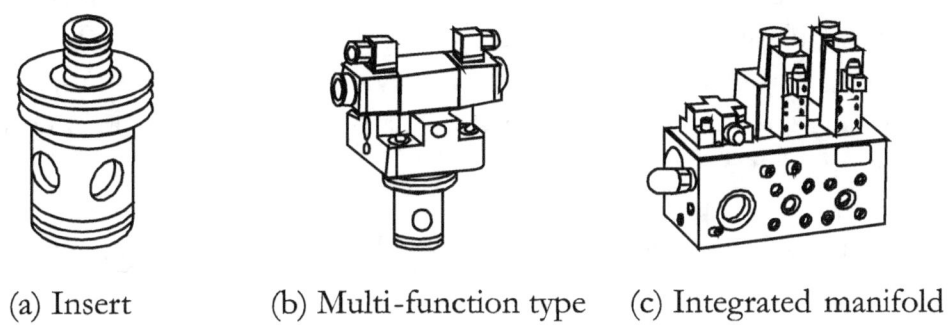

(a) Insert   (b) Multi-function type   (c) Integrated manifold

Figure 1.2 | Different types of hydraulic cartridge valves

The primary cartridge valve is a single-function valve that realizes the fundamental 2/2-way directional (or logic) valve function.

Figure 1.2(a) shows the schematic diagram of the insert in the primary cartridge valve. Such a valve is the basic unit for the more sophisticated cartridge valve system.

Figure 1.2(b) shows the schematic diagram of a multifunction cartridge valve. Such a valve can be constructed by combining the basic cartridge element with an appropriate control cover.

The multi-function design of the cartridge valve, including check (non-return), directional, pressure, and flow functions, can be achieved by combining the cartridge element with the control cover.

The control cover can also contain shuttle valves, pilot-operated check valves, pressure relief valves, and proportional pressure relief valves. It may also be provided with a stroke limiter with a manual adjustment to limit the poppet lift and control the flow rate through the valve.

A control cover can be connected to a solenoid-operated 3-way or 4-way directional control pilot valve with proper interfacing to achieve advanced check, directional, flow, and pressure control functions.

Combining different types of cartridge valves in the manifold block can integrate a partial or complete hydraulic system into a single manifold.

Early cartridge valves were used as screw-in or slip-in components, primarily for basic pressure regulation, flow control, and check-valve functions. The advent of 2/2-way valves, commonly known as logic valves, enabled a wider range of manifold applications, delivering enhanced performance under elevated pressure and minimizing leakage.

Figure 1.2(c) shows the schematic diagram of the integrated manifold with many cartridge valves.

Manifold block construction eliminates many interconnecting lines between components, virtually eliminating potential leakage points and related fluid waste. Cartridges of the correct sizes can be conveniently installed in an optimally designed manifold to reduce space requirements and overall costs.

General Characteristics of Cartridge Valves

The standard cartridge valves are available in a broad range of sizes. High-flow non-standard cartridge valves are available for specialized fluid power applications that need higher performance.

It may be noted that the cavities in cartridge valve manifolds and their mounting pattern are specified under the ISO 7368 standard.

Typically, the standard manifold systems contain valves with nominal bores of 16, 25, 32, 40, 50, 63, 80, and 100 mm, and the non-standard manifold systems contain valves with nominal bores up to 160 mm. The cartridge valves can operate at pressures up to 420 bar [6000 psi] and with flows up to 25000 lpm [6600 gpm].

The control covers come with inch- or metric-threaded pilot valve mounting bolts, gauge ports, and orifice plugs.

Chapter 2 | Constructional Features and Circuits of Single-function Cartridge Valves

A single-function cartridge valve is designed as a logic element for insertion into the cavity of a compact manifold block and sealed utilizing a cover plate to form part of an integrated hydraulic circuit for directional control. Many cartridge valve manufacturers make cartridges that fit the standard cavities. A cartridge insert can be removed from its manifold with the standard cavity and replaced with another one from a different manufacturer.

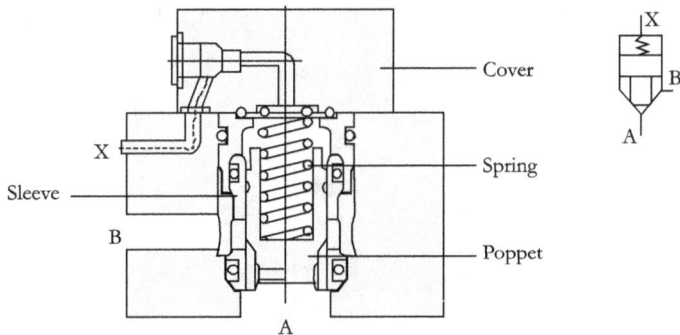

Figure 2.1 | A cutaway view of a basic cartridge valve

The single-function cartridge valve, shown in Figure 2.1, consists of a spring-biased cartridge element, a cover, and necessary ports. The cartridge comprises a poppet, a sleeve, and a spring chamber (A_P) to contain a closing spring. Drilled holes in the manifold connect the main ports A and B to other cartridges or the operating hydraulic system. An internal pilot orifice is provided between port A and the spring chamber AP to ensure shock-free application of control signals.

A cartridge valve poppet (cone) is optionally provided with a damping nose; some come with notches (orifices) for flow control. The damping nose reduces shock, vibration, and noise in the associated hydraulic system, improving system performance.

The spring frequently holds the poppet in its closed position against a leak-proof seat. The control spring provides a biasing force to close the poppet. Manufacturers typically offer springs with strengths of 1.0 bar, 1.9 bar, and 3.8 bar (14.5 psi, 28 psi, and 55 psi).

A basic cover fitted above the insert acts as a control interface for the valve poppet through a control passage with an orifice. More advanced covers may contain built-in poppet or shuttle valves and stroke-limiting options. They can also be used to mount directional poppet or spool valves. Depending on the type of cover, the cartridge valve can perform directional, throttle, or pressure functions, or combinations of these, as we will see in subsequent chapters.

A cartridge valve consists of two working (operational) ports, A and B, and the pilot port X. The critical areas of the poppet are A_A, A_B, and A_X (or A_{AP}), corresponding to the effective areas of ports A, B, and X, respectively. These critical areas are shown in Figure 2.2. The ports must conform to NPT, SAE, or other standards.

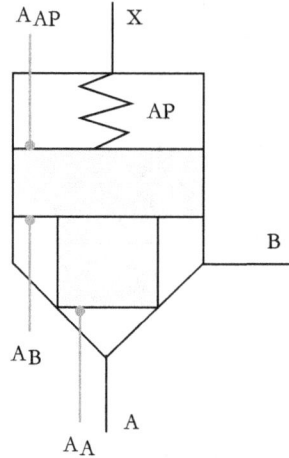

Figure 2.2 | Cartridge valve symbol with critical areas

The position of the poppet is decided by the balance of forces acting on both sides of the poppet. The closing force (F_c), due to the pressure acting on the area A_X and the spring, tends to close the valve, whereas the opening force (F_o), owing to the pressure acting on the area A_A and A_B, tends to open the valve. Therefore, the flow path between port A and port B can be controlled hydraulically by a pilot pressure applied to port X.

In general, valve poppets may have different geometric shapes and operating area ratios to optimize control. The cartridge valve is usually constructed so that the area AX equals the sum of the areas A_A and A_B. Typically, the area ratio between A_A and A_X ($A_A:A_X$) can be 1:2, 1:1.6, 1:1.1, 1:1.05, 1:1, etc. For instance, if the area ratio is 1:2, area A_X is twice the size of area A_A, and area A_A is the same size as area A_B. If the area ratio is 1:1, area A_X equals area A_A, and area A_B equals zero. Note that the valve with an area ratio of 1:1 is used for pressure control.

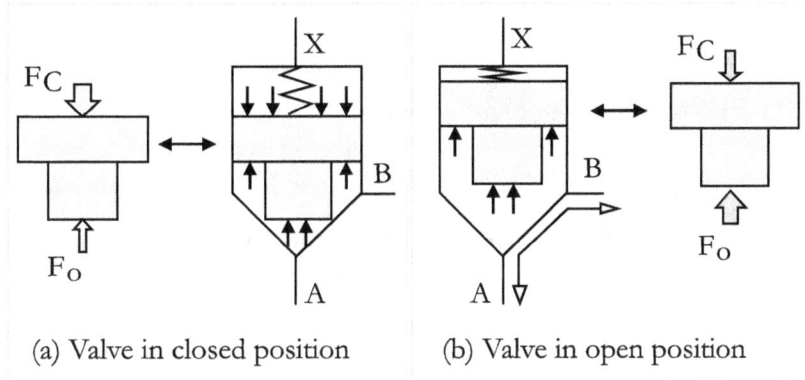

(a) Valve in closed position   (b) Valve in open position

Figure 2.3 | Illustrating the operation of a cartridge valve

Figure 2.3 illustrates the operation of the cartridge valve. If the pilot signal is present at port X, the closing force F_c becomes greater than the opening force F_o. The poppet is, therefore, pushed onto the valve seat, and the valve remains tightly closed, as shown in Figure 2.3(a). In this case, the working connection from port A to port B is shut off.

If the pilot signal X is not present, the closing force Fc is less than the opening force Fo; the poppet is forced away from its seat, and the valve opens, as shown in Figure 2.3(b). In this case, the working connections between port A and port B are open, allowing fluid to flow in either direction, from port A to port B or from port B to port A. A manually operated or solenoid-operated directional control valve can switch the pilot line.

Material, Cartridge Valves

High-pressure cartridge valves and manifolds (>210 bar) use leaded steel that is zinc-plated and treated with yellow trivalent chrome for optimal performance and durability.

Medium-pressure valves (<210 bar) can be made from high-strength aluminum. Solenoid-operated directional valves use cast iron that is zinc-plated with yellow trivalent chrome.

Seal Materials, Cartridge Valves
Acrylonitrile butadiene (NBR) can be used as a seal for O-rings in cartridge valves. This material works well with mineral fluids, water-in-oil emulsions, and water-glycol fluids. It can withstand temperatures ranging from -30°C to +100°C (-22°F to +212°F). Strengthened PTFE is suitable for backup rings and slide rings. To ensure optimal performance, cartridge valves operating at higher temperatures must use special FPM (FKM) seals.

Filtration Requirements, Cartridge Valves
Manufacturers recommend a fluid cleanliness class for cartridge valves to prevent component malfunction and increase service life. A typical recommendation for fluid cleanliness, per ISO 4406, is 20/18/15.

Spring and spring rate
Poppets can be combined with different springs to provide different cracking pressures. Typical opening pressures are 0.1, 0.5, 1.6, 2.5, and 4.0 bar. (1.45, 7.25, 23.2, 36.25, and 58 psi)

Basic Types of Cartridge Valves
Depending on how a cartridge is fixed in its cavity, there are two basic types of cartridge valves. They are: (1) screw-in type and (2) slip-in type. The classification is shown in Figure 2.4.

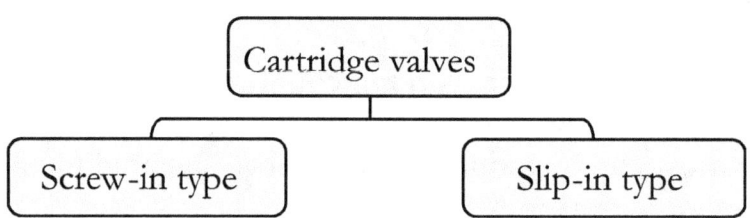

Figure 2.4 | Basic types of cartridge valves

The screw-in type cartridge valve is configured when its element mates with the threads in its cavity. The screw-in type cartridge valves are typically low-flow valves with flow rates < 75 lpm [<20 gpm]. A control cover bolted to the manifold retains the slip-in cartridge in the cavity. Slip-in cartridge valves are associated with relatively high flows, i.e., flow rates >150 lpm [>40 gpm]. They target more efficient, faster, and more compact hydraulic systems.

Symbols of Cartridge Valves

Figure 2.5 shows some control components used in cartridge valve systems and their variants.

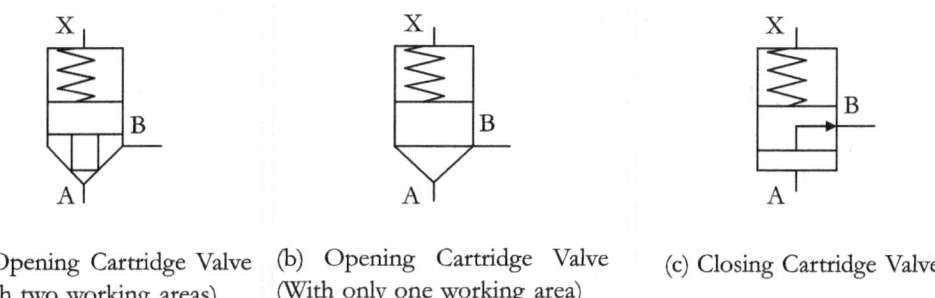

(a) Opening Cartridge Valve (With two working areas) (b) Opening Cartridge Valve (With only one working area) (c) Closing Cartridge Valve

Figure 2.5 | Symbols of cartridge valves

Figure 2.5(a) shows the symbol of an opening cartridge valve with two working areas, A_A and A_B. Figure 2.5(b) shows the symbol of an opening cartridge valve with only working areas A_A ($A_B = 0$). Figure 2.5(c) shows a closing cartridge valve. The valve remains open when a signal is applied to the pilot port.

Symbols – Control Covers

A control cover is bolted to the manifold leak-free to retain the associated insert. Various control covers with standard ports are available. Typically, the ports are designated as X, AP, Z1, Z2, Y, P, A, B, and T.

Port AP is to be connected to the cartridge valve's spring chamber. The signal to control the cartridge's opening or closing is sent to port X.

Ports Z1 and Z2 are also used to apply remote pilot signals.

Y is the drain port.

Ports P, A, B, and T interface with a 4-way directional control valve.

Figure 2.6 shows the basic directional control cover used in cartridge valve systems and their variants.

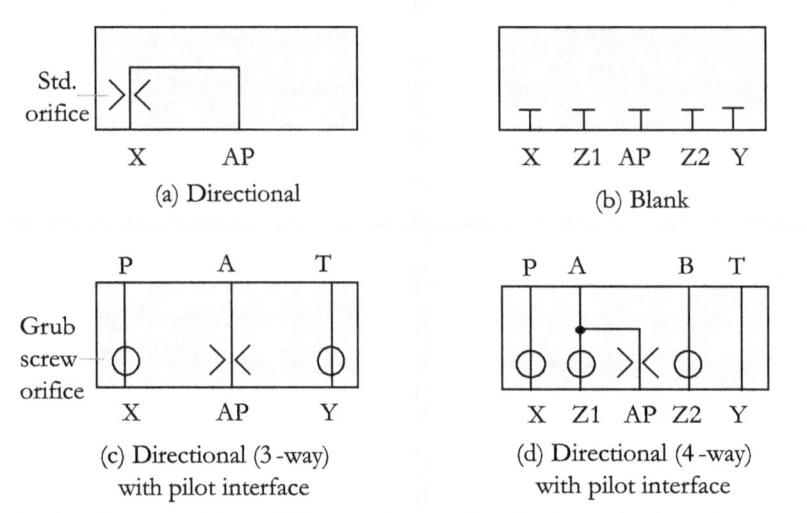

Figure 2.6 | Symbols of basic covers of cartridge valves

A cartridge cover must operate reliably even under extreme loading conditions. It must be selected with a pressure rating suitable for the seal material used in the cover and the highest load expected in the associated system. The cover must also allow for maximum flow and the lowest pressure drop. Control covers that integrate check valves and shuttle valves, and interfaces for subplate pilot valves, must support a compact system design, even for complex functions.

Cartridge Valve Basic Circuits
Cartridge valve circuits are very versatile when set up with various control valves. Because the cartridge valve is a two-position valve, it can control the flow between only two major flow points in the system. Cartridge valves for directional and check functions are essentially hydraulically piloted check valves.

Control options for cartridge valve systems include single- or multiple-pilot arrangements, flow restrictors, and solenoid-controlled, pilot-operated directional control valves. Combining cartridge assemblies can provide directional, check, and/or flow restrictor functions for normal flow rates up to 2800 lpm (740 gpm) per cartridge.

Some simple, self-explanatory circuits of cartridge valves are given in the following sections.

Example 2.1 | A Circuit with an Opening Cartridge Valve Having Two Working Areas for A to B Directional Control

Figure 2.7 shows the basic circuit of an opening cartridge valve with two operating areas, A_A and A_B. A pump is connected to port A of the valve, and a hydraulic motor is connected to port B. The pilot signal (X) is controlled externally using a 3/2-way valve. Figure 2.8 shows multiple circuit configurations for directional control from port A to port B.

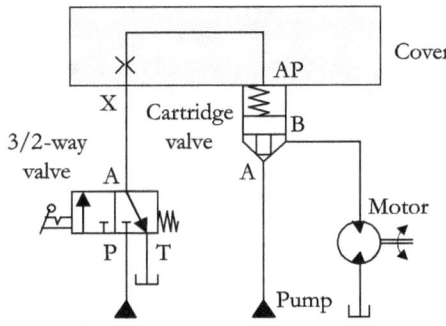

Figure 2.7 | A circuit with an opening cartridge valve for A to B control

If a signal is present at the spring chamber port AP (the 3/2-way valve is actuated), the cartridge valve remains closed, and flow from port A to port B is blocked, as shown in Figure 2.8(a). If the pilot signal is not present at port AP (the 3/2-way valve is released), flow from port A to port B is possible, as shown in Figure 2.8(b).

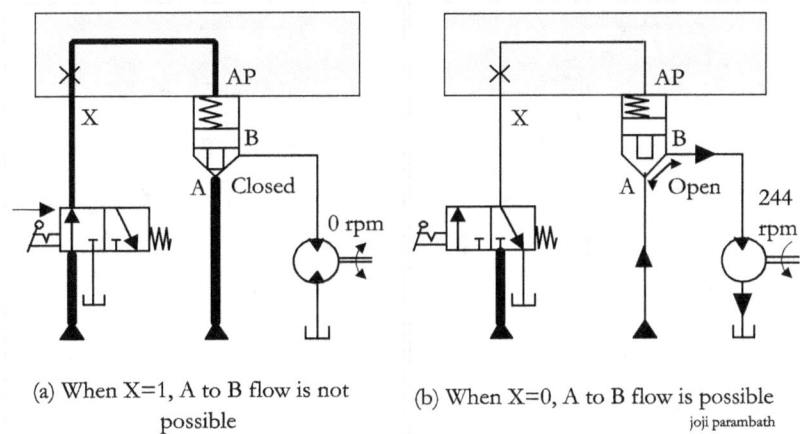

(a) When X=1, A to B flow is not possible

(b) When X=0, A to B flow is possible

Figure 2.8 | Multiple positions of the circuit for A to B control

Example 2.2 | A Circuit with an Opening Cartridge Valve Having Two Working Areas for B to A Directional Control

Figure 2.9 shows another directional control circuit of an opening cartridge valve with two working areas. A pump is connected to port B of the cartridge valve, and a hydraulic motor is connected to port A of the cartridge valve. The pilot signal (X) is controlled externally using a 3/2-way valve. The critical operating positions of the circuit with an opening cartridge valve are given in Figure 2.10.

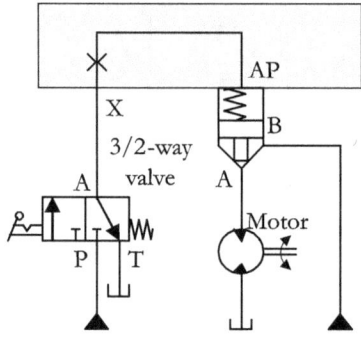

Figure 2.9 | A circuit with an opening cartridge valve for B to A control

If a signal is present at the spring chamber port AP, the valve remains closed, and flow from port B to port A is blocked, as shown in Figure 2.10(a). If the pilot signal is not present at port AP, flow from port B to port A is through, as shown in Figure 2.10(b).

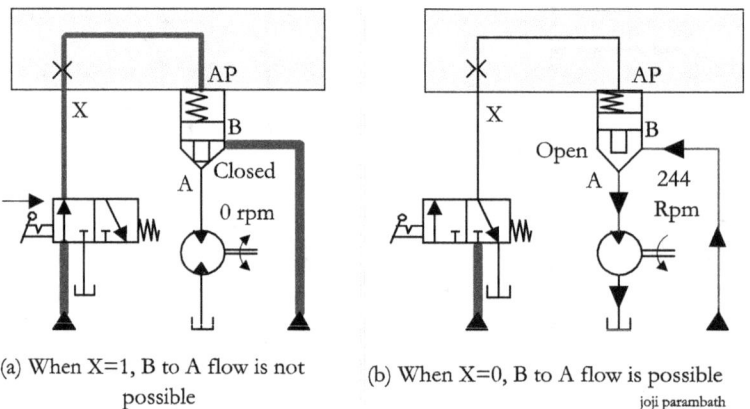

(a) When X=1, B to A flow is not possible

(b) When X=0, B to A flow is possible

Figure 2.10 | Multiple positions of the circuit with an opening cartridge valve for B to A control

Example 2.3 | A Circuit with a Closing Cartridge Valve

Figure 2.11 demonstrates a basic circuit of a closing cartridge valve. A pump is connected to port A of the cartridge valve, and a hydraulic motor is connected to port B. The pilot signal (X) is controlled externally using a 3/2-way valve. Figure 2.12 shows the critical operating positions of the circuit with a closing cartridge valve.

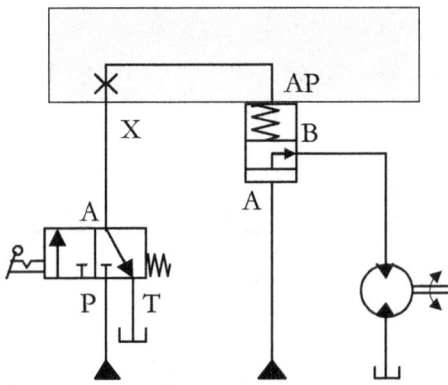

Figure 2.11 | A basic circuit with a closing cartridge valve

If the signal is not present at the spring chamber port AP, the valve remains closed, and flow from port A to port B is blocked, as shown in Figure 2.12(a). If the pilot signal is present at port AP, flow from port A to port B is through, as shown in Figure 2.12(b).

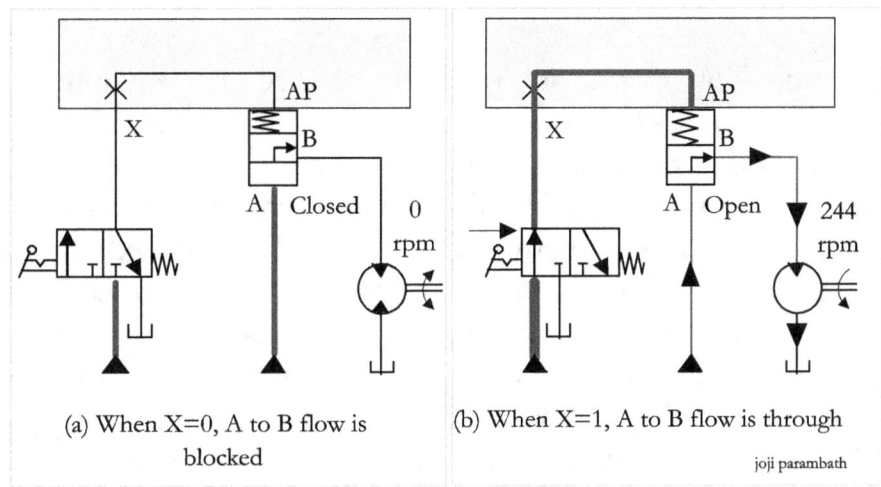

(a) When X=0, A to B flow is blocked

(b) When X=1, A to B flow is through

Figure 2.12 | Multiple positions of the circuit with a closing cartridge valve

Passive and Active Logics

A logic assembly with only one control area in the spring chamber of a cartridge valve is termed passive logic. In contrast, a logic assembly having a differential spool with at least one additional control area is termed active logic. The pilot pressure in the additional control area can keep the active logic open even when ports A and B are not pressurized. Chapter 8 explains how active logic works.

Cartridge Cover Dimensions

Figure 2.13 shows a typical cartridge valve mounting pattern for sizes 16 to 63 and cover dimensions for size 32, as per ISO 7368.

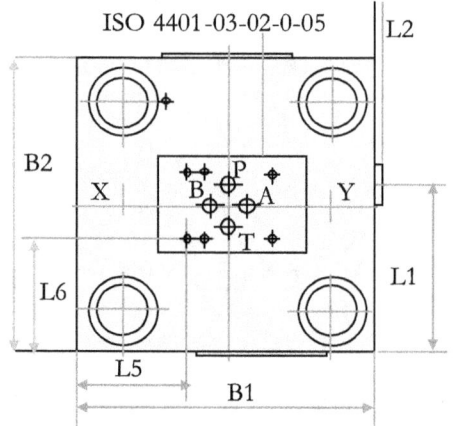

Cover dimensions

Size	32
B1 mm (in)	102 (4.02)
B2 mm (in)	102 (4.02)
L1 mm (in)	61.3 (2.4)
L2 mm (in)	3.5 (0.14)
L5 mm (in)	32 (1.26)
L6 mm (in)	34.65 (1.36)
ISO 7368 mounting pattern	09-5-1-16

Figure 2.13 | Cartridge cover dimensions

What Next....

Looking forward, there is the potential to enhance the capabilities of single-function cartridge valves by combining them with suitable covers. These covers come in many standard designs with unique functions for controlling direction, pressure, and flow. By combining the various covers, operating units, and connections within the block, many functions can be achieved, including direct control, non-return, hydraulically piloted non-return, pressure control, flow rate regulation, and combinations of these functions. This opens countless possibilities for using cartridge valves for directional, pressure, and flow control. With such versatility and superior performance, integrating cartridge valves in various applications is becoming more common. The upcoming chapters will provide valuable insights into these control functions, enhancing your understanding.

Chapter 3 | Constructional Features of Multifunction Cartridge Valves and Circuits for Check Function

Combining the valve insert with a suitable cover can integrate more features into a single cartridge valve. Many standard covers are designed with different functions to allow flow in one direction while blocking it in the other. Accordingly, cartridge valves can be used to realize the check function. The multifunction logic valves enable compact, integrated hydraulic systems that reduce manifold dimensions and the number of ports while increasing flow-path efficiency. These valve configurations are lower-cost than conventional hydraulic valves.

Cartridge valve circuits are very versatile when set up with various control valves. Because the cartridge valve is a two-position valve, it can control the flow between two major flow points in the system. Cartridge valves for check function are primarily used for unidirectional flow switching applications. The following sections give some simple, self-explanatory circuits of cartridge valves.

Basic Cartridge Valve Circuit for the Check Function

Figure 3.1 shows a partial circuit representation of a typical single-cartridge arrangement with an insert and a basic control cover for the check function. The insert consists of main ports A and B, and the control cover consists of an orifice, port AP, and port X. Port AP of the cover is connected to the spring chamber of the insert. Port X should be connected to port B for the check function. The valve provides a check function by allowing free flow from A to B while blocking flow from B to A.

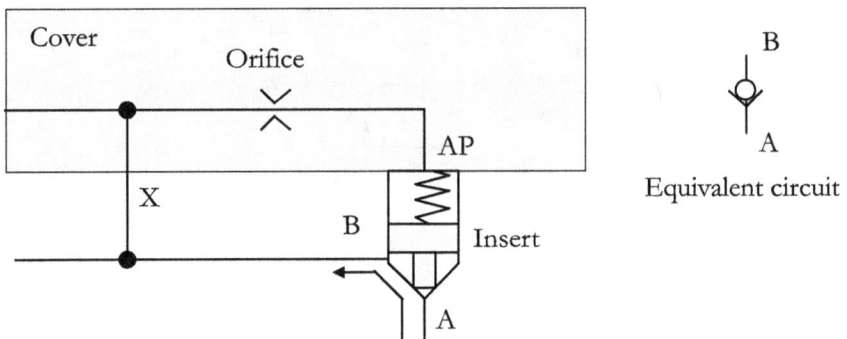

Figure 3.1 | A partial cartridge valve circuit configuration for a check function

Example 3.1 | Check Function with Internal Piloting

The pilot signal to control the cartridge valve can be generated internally by direct pressure from either port A or port B. Figure 3.2 shows the internal piloting from port B.

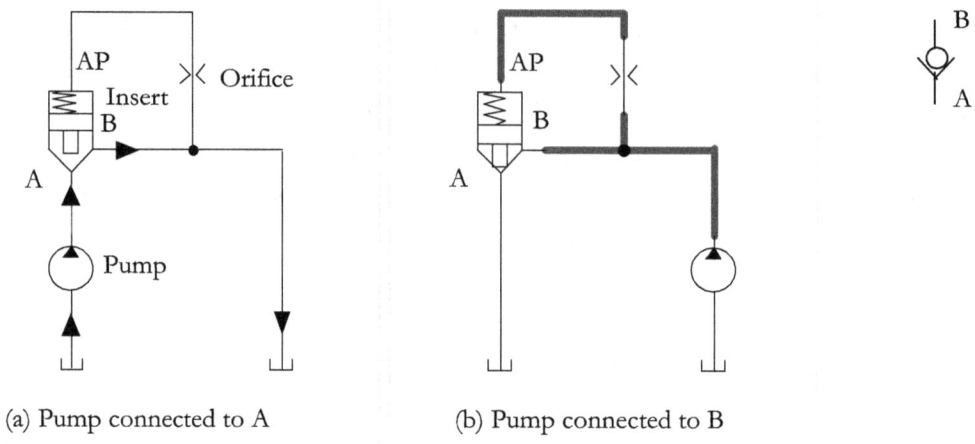

(a) Pump connected to A

(b) Pump connected to B

Figure 3.2 | Internal piloting (B to AP)

Basic Cartridge Valve for the Direct-acting Check Function

Figure 3.3 shows a basic cartridge valve configuration for the direct-acting check function. The free-flow direction is from A to B. Port B is connected to the spring chamber through a drilled hole in the poppet.

The pressure at port B also acts on the full poppet area, A_{AP}, where it assists the spring in holding the poppet tightly against the valve seat. Therefore, the flow in the opposite direction, B to A, is blocked.

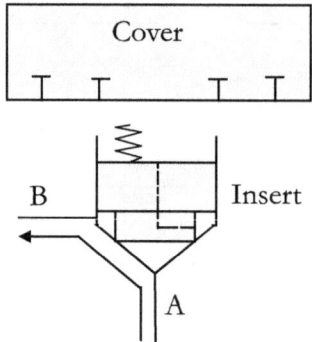

Figure 3.3 | A basic cartridge valve for the direct-acting check function

Cartridge-type Check Valve for External Piloting

With the internal pilot port blocked, as shown in Figure 3.4, external pilot pressure can be applied through pilot port Z1 to close the valve. Without a pilot signal to port Z1, the pressure applied to port A or port B opens the valve. That is, the cartridge configuration acts like a check valve.

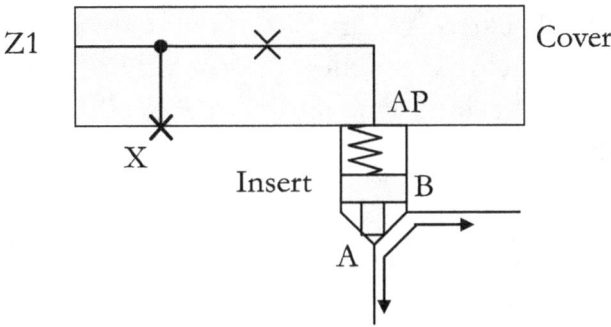

Figure 3.4 | A cartridge configuration for a check valve with external piloting

A Cartridge Configuration for Pilot-operated Check Valve

Figure 3.5 shows a cartridge valve configuration with a multi-function cover, a pilot-operated check valve, and ports X, Z1, AP, and Y. Port X is connected to port B, and port Y is to be connected to a tank.

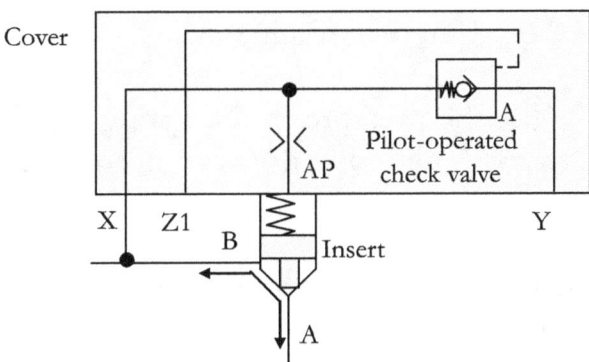

Figure 3.5 | A multi-function cartridge configuration with a pilot-operated check valve

Without a pilot signal at Z1, pressure at port B directly pressurizes the spring chamber AP. Therefore, flow from port B to port A is not permitted. However, flow from port A to port B is permitted.

When a pilot signal is applied to port Z1, pressure in the spring chamber AP is relieved through port Y, permitting flow from port A to port B and port B to port A. Typically, pressure at port Z1 must be at least 30% of load pressure, including any pressure intensification that may occur.

Example 3.2 | External Piloting using 3/2-DC valve

Figure 3.6 shows a simple cartridge circuit with an external pilot feature using an electrohydraulic 3/2-way single solenoid pilot valve. Port B of the insert is connected to port P of the pilot valve, and port A of the pilot valve is connected to port X of the insert.

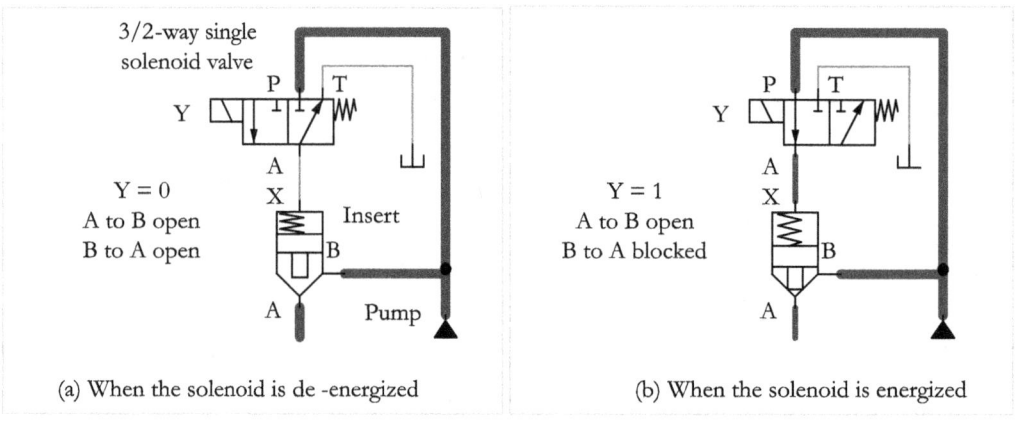

Figure 3.6 | External piloting using 3/2-DC valve

As shown in Figure 3.6(a), the pressure in the spring chamber is relieved when the solenoid is de-energized. Therefore, flow can occur from port A to port B or from port B to port A.

On the other hand, when the solenoid is energized, as shown in Figure 3.6(b), the pump flow is directed to the spring chamber of the cartridge valve. Therefore, the flow from port B to port A is blocked. However, the flow from port A to port B is possible if the pump is connected to port A.

Note: For simplicity, the cartridge valve is shown to be connected directly to the 3/2-way pilot valve. However, it must be connected to the pilot valve through a proper cover with matching interfacing openings for ports P, A, and T, as described in the following chapters.

Chapter 4 | Control Covers and Circuits for Directional Controls

Directional control can be realized using a single-function cartridge valve or a group of cartridge valves. The cartridge valve covers offer control options, including single- or multiple-pilot arrangements, flow restrictors, and mounting interfaces for solenoid-operated directional control valves. A cartridge valve for the basic 2-way directional function is an externally piloted insert with a basic control cover. More extensive directional control requires the assembly of multi-function cartridge valves, as in a bridge circuit. Typical examples of directional control are given in the following sections.

It may be noted that a cartridge valve for directional control typically can handle flow up to 2800 lpm (740 gpm).

Example 4.1 | 2-way, 2-position Function

Figure 4.1 shows a cartridge valve assembly with an insert and a basic control cover. The cartridge insert with manifold assembly has main ports A and B. The control cover typically has port AP for connection to the insert's spring chamber and pilot pressure ports X and Z1. For the 2-way directional control, internal port X should be blocked, and the pilot signal is given externally through port Z1.

Pressurizing port Z1 from a remote source will block the flow from port A to port B. However, relieving the pilot pressure in the spring chamber will permit the flow from port A to port B or from port B to port A. Therefore, this arrangement can provide directional control of the flow from port A to port B or from port B to port A.

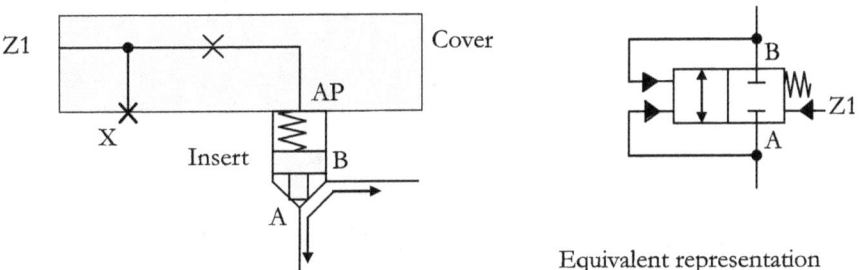

Equivalent representation

Figure 4.1 | A cartridge valve configuration for 2-way 2-position control

Example 4.2 | An Electro-hydraulic Circuit for the ON/OFF Control of a Hydraulic Motor Using a Cartridge Valve Employing External Piloting with a 3/2-way Solenoid Valve

Figure 4.2 shows an electro-hydraulic circuit for the ON/OFF control of a hydraulic motor using a cartridge valve with a control cover. The cover has a mounting interface for a directional control pilot valve. A pump supplies the necessary fluid at the required pressure. A 3/2-way normally open (NO) type solenoid valve is used to pilot the cartridge valve externally. The control cover interfaces the 3/2-DC valve and the cartridge insert. Multiple circuit positions are also shown in Figure 4.3.

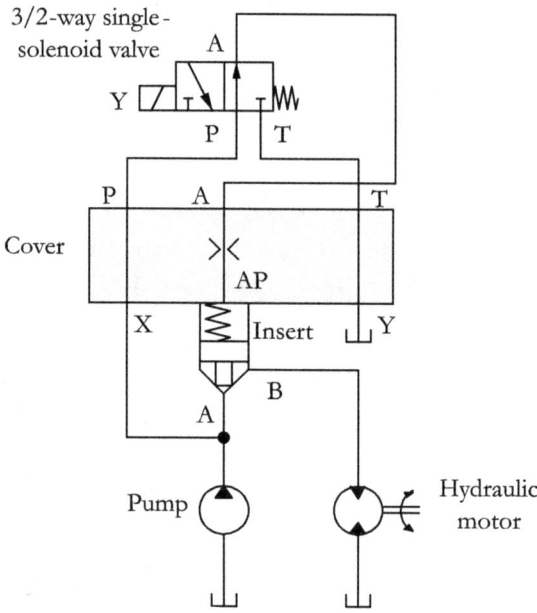

Figure 4.2 | An electro-hydraulic cartridge circuit employing external piloting for the control of a hydraulic motor (Example 4.2)

Figure 4.3(a) shows the electro-hydraulic circuit when pushbutton PB is released, and the 3/2-way pilot valve is in its normal position. In this circuit position, the pump flow is directed to the cartridge valve's pilot port AP through the 3/2-way valve, which is tightly closed. As a result, the pump flow cannot pass through the cartridge valve, which prevents the hydraulic motor from running. The pilot valve, with standard locations and port and hole patterns for mounting bolts and locating pins, can be installed directly on the cover, with matching hole patterns for ports, mounting bolts, and locating pins.

Figure 4.3(b) shows the circuit's position when pushbutton PB is pressed, and the 3/2-way valve is actuated. In this circuit position, the pilot port pressure is relieved to the tank through the 3/2-way pilot valve, and the cartridge valve remains open when pressure is applied to port A. The pump flow can pass through the cartridge valve and drive the hydraulic motor.

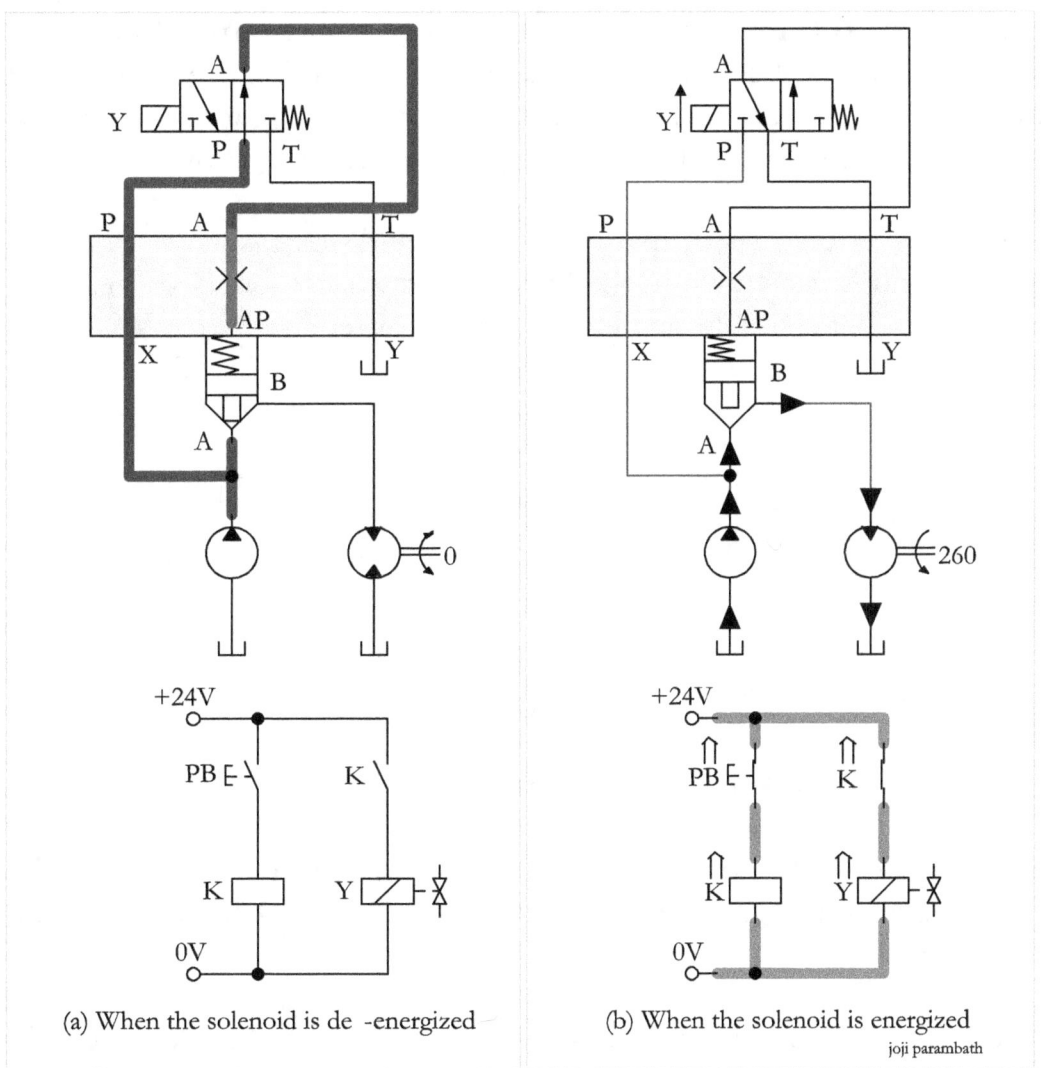

Figure 4.3 | A circuit for the ON/OFF control of a hydraulic motor using a cartridge valve employing external piloting (Example 4.2)

A properly sized and correctly positioned orifice in the cartridge valve's control cover enables precise regulation of pressure and flow rate, increasing efficiency and reliability. Typical orifice sizes are 0.8, 0.9, 1.0, 1.2, and 1.4 mm.

Multifunction Cartridge Valves

Combining the valve insert with a suitable cover allows more features to be integrated into a single cartridge valve. Many standard covers have different functions for controlling direction, pressure, and flow. Accordingly, cartridge valves can realize these functions.

Control Covers to Interface with 4-way Directional Control Valves

Cartridge valve control covers are available with mounting interfaces for single-solenoid or double-solenoid directional control valves. Figure 4.4 shows control covers for interfacing the insert of cartridge valves with directional control valves.

The control cover consists of many flow paths and ports, such as P, A, B, T, X, Z1, AP, Z2, and Y, as shown in the Figure. Port Z1 can be used for the remote hydraulic control and should be blocked when not required. Port Y is usually the tank port.

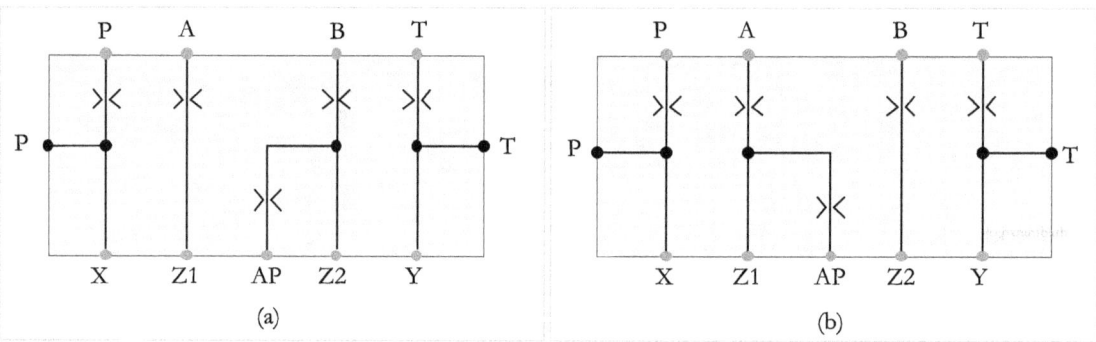

Figure 4.4 | Covers for cartridge valves as directional control interfaces

Figure 4.4(a) shows the control cover that connects port B to the pilot line AP internally. Figure 4.4(b) shows the control cover that connects port A to the pilot line AP internally.

The following sections show that each control cover can be used with the cartridge insert and 4-way valve to configure a cartridge valve system. The internal pilot port X is connected internally to port B of the valve insert, and the pilot port Y is connected to the drain.

Note: Appendix 3 details the ISO mounting configurations of 4-port directional control valves.

Example 4.3 | ON/OFF Cartridge Valve, Dumping While Energized

Develop an equivalent ON/OFF valve using a cartridge valve with a suitable control cover and a 4-way solenoid-operated directional control valve. The valve should dump when the solenoid is energized.

Solution

Figure 4.5 shows multiple positions of the circuit. The circuit shown in Figure 4.5(b) allows the cartridge valve to dump in the solenoid's energized state, and the circuit shown in Figure 4.5(a) allows the cartridge valve to block the flow in the solenoid's de-energized state.

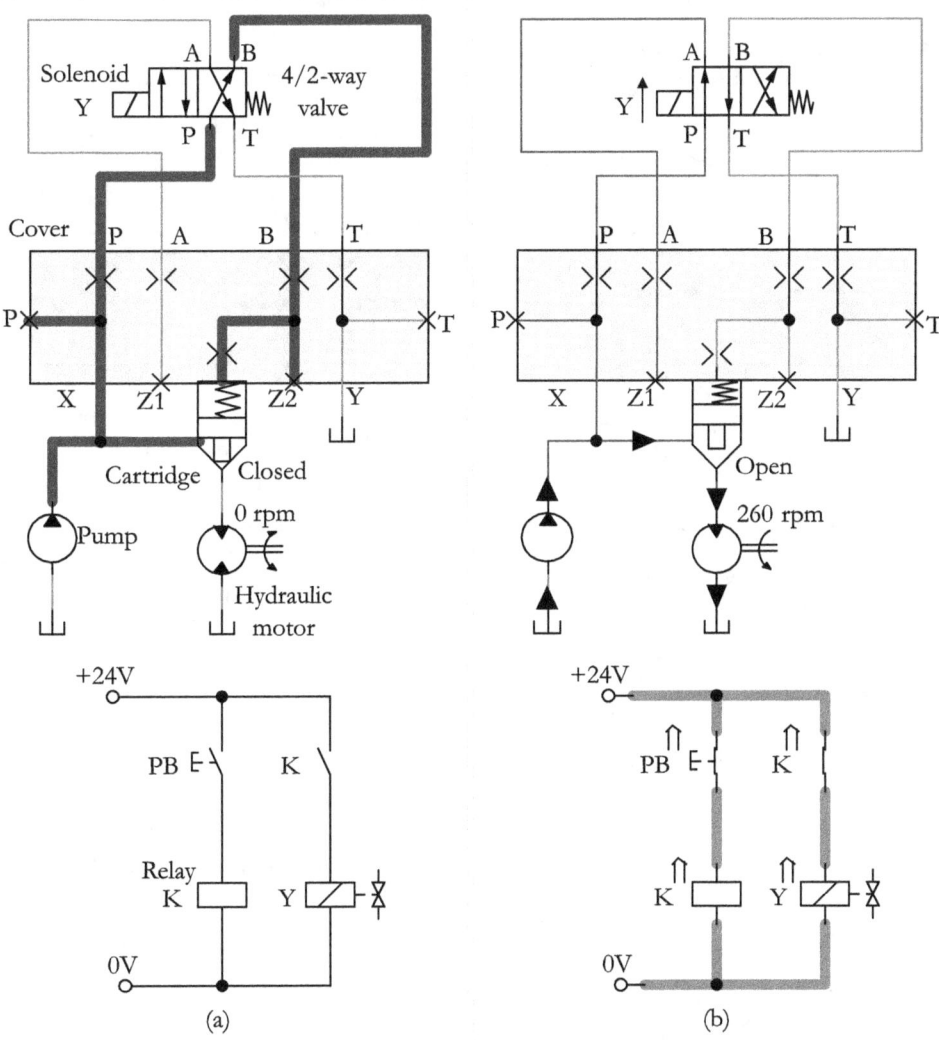

Figure 4.5 | An ON/OFF cartridge valve circuit, dumping while energized

Example 4.4 | ON/OFF Cartridge Valve, Dumping While De-energized
Develop an equivalent ON/OFF valve using a cartridge valve with a suitable control cover and a 4-way solenoid-operated directional control valve. The valve should dump when the solenoid is de-energized.

Solution

Figure 4.6 shows multiple positions of the circuit. The circuit shown in Figure 4.6(a) allows the cartridge valve to dump in the solenoid's de-energized state, and the circuit shown in Figure 4.6(b) allows the cartridge valve to block the flow in the solenoid's energized state.

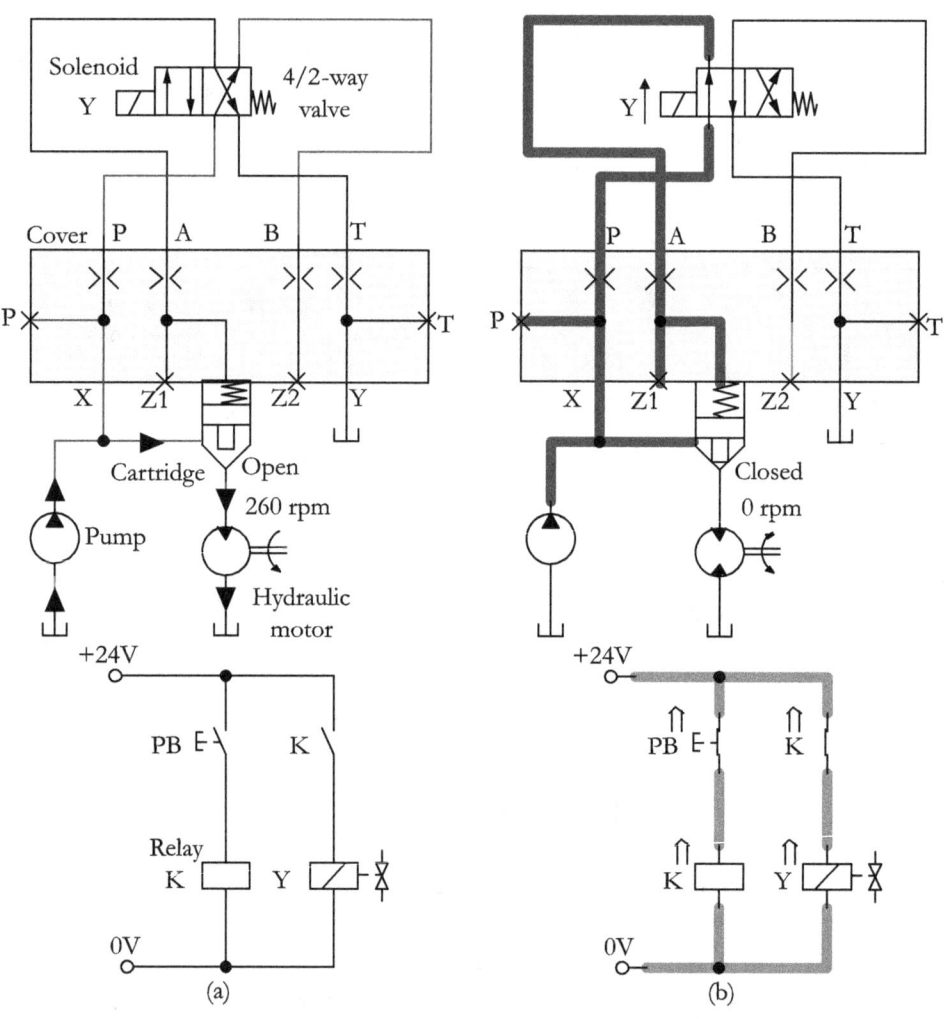

Figure 4.6 | An ON/OFF cartridge valve circuit dumping while de-energized

Control Cover with a Shuttle Valve

Figure 4.7 shows the circuit diagram of a cartridge valve with a control cover. The control cover incorporates a shuttle valve with ports P, A, T, Z1, X, AP, and Y. The shuttle valve directs higher pressure to the X port or Z1 port (typically through an energized 3/2- or 4/2-way solenoid valve) to the poppet insert spring area (AP), closing the cartridge valve.

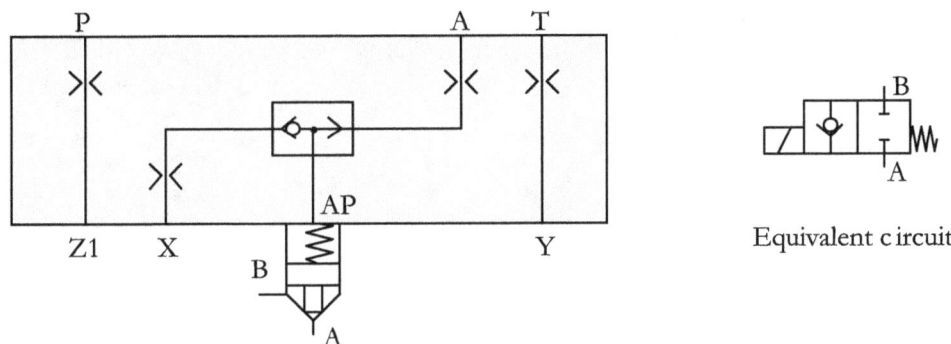

Figure 4.7 | A control cover with a shuttle valve

In the de-energized state of the solenoid, the control port AP is typically connected to the tank (T), keeping the cartridge valve normally open.

Example 4.5 | Pilot Shuttle Control

A pump drives a hydraulic motor through a cartridge valve control arrangement. The cartridge valve system is designed to have two switching positions realized through a 3/2-way solenoid valve. In the normal position of the 3/2-way valve, the main flow through the cartridge flow path should be blocked. In the actuated position of the 3/2-way valve, the cartridge system should permit the flow. Develop a hydraulic circuit with the pump connected to port A and the hydraulic motor connected to port B.

Solution

Two positions of the cartridge valve circuit for controlling the hydraulic motor powered by a pump are given in Figure 4.8. The pump is connected to port A of the insert, and the motor to port B.

The cartridge valve system's switching positions are achieved using a 4/2-way single-solenoid pilot valve. Port X of the control cover is connected to port B of the insert, and port Z1 is connected to port A of the insert.

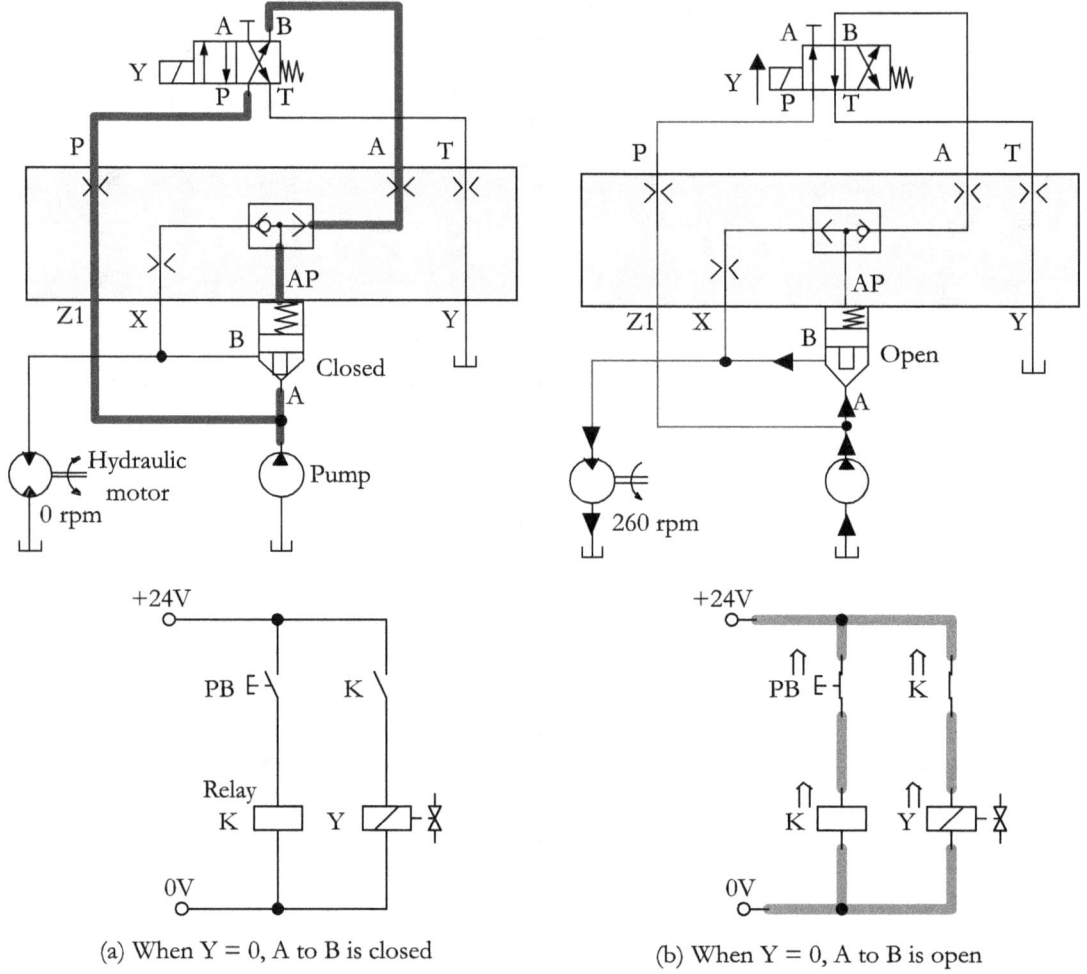

(a) When Y = 0, A to B is closed   (b) When Y = 0, A to B is open

Figure 4.8 | Multiple positions of the cartridge valve circuit (Example 4.5)

Figure 4.8(a) shows the circuit when the solenoid is de-energized. In this circuit position, the pump flow energizes the spring chamber through the 4/2-way valve and the shuttle valve.

Therefore, the cartridge valve remains closed, and the pump cannot drive the hydraulic motor.

Figure 4.8(b) shows the circuit when the solenoid is energized. In this circuit position, the spring chamber is relieved to the tank.

Therefore, the cartridge valve opens, and the pump drives the hydraulic motor.

4-way Directional Function Using Multiple Cartridge Valves

Many working positions of 4-way valves can be realized by assembling multiple cartridge valves in a bridge arrangement. Some switching positions of the 4-way valve can be achieved by energizing or de-energizing the solenoids, as per the self-explanatory control schematic in Table 4.1. A bridge circuit configuration is explained in Example 4.6.

Table 4.1 | Switching states of the solenoids for obtaining various positions of a 4-way DC valve

Solenoid	Open-center	Float-center
Y1	ON	ON
Y2	ON	OFF
Y3	ON	OFF
Y4	ON	ON

Example 4.6 | Directional Control of a Double-acting Hydraulic Cylinder Using Cartridge Valves in Bridge Circuit Arrangement

Develop an electrohydraulic circuit with four cartridge valves in a bridge circuit arrangement and four 3/2-way solenoid-operated directional control valves to realize the switching positions to control a double-acting cylinder's forward and return strokes.

Solution

Figure 4.9(a) shows the hydraulic circuit for directional control of the double-acting hydraulic cylinder using four cartridge valves, CV1, CV2, CV3, and CV4, in the bridge-circuit arrangement. The pilot lines of the cartridge valves are controlled by the respective 3/2-way solenoid coils Y1, Y2, Y3, and Y4.

The simplified electrical circuit for controlling the solenoid coils is given in Figure 4.9(b).

Let us assume pump 1 supplies the main flow and pump 2 supplies the control flow. Multiple circuit positions for the forward and return strokes are given in the following circuits.

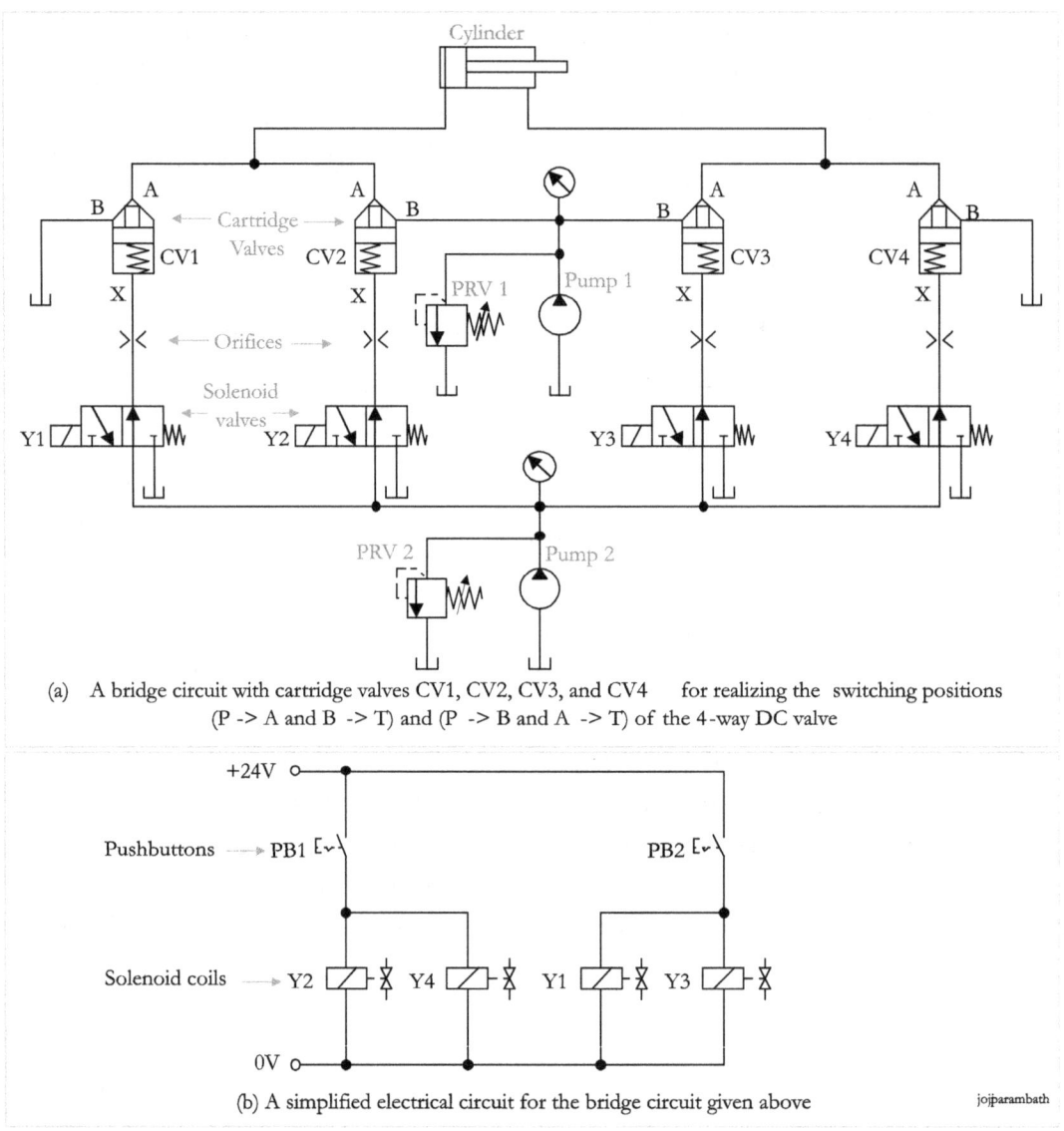

(a) A bridge circuit with cartridge valves CV1, CV2, CV3, and CV4 for realizing the switching positions (P -> A and B -> T) and (P -> B and A -> T) of the 4-way DC valve

(b) A simplified electrical circuit for the bridge circuit given above

Figure 4.9 | The electro-hydraulic circuit with cartridge valves in bridge configuration and the associated electrical circuit (Example 4.6)

The Position of the Circuit When Pushbutton PB1 is Pressed

Figure 4.10 shows the position of the electro-hydraulic circuit when pushbutton PB1 is pressed for the forward motion of the double-acting cylinder.

The hydraulic part of the circuit is given in Figure 4.10(a), and the electrical part is given in Figure 4.10(b).

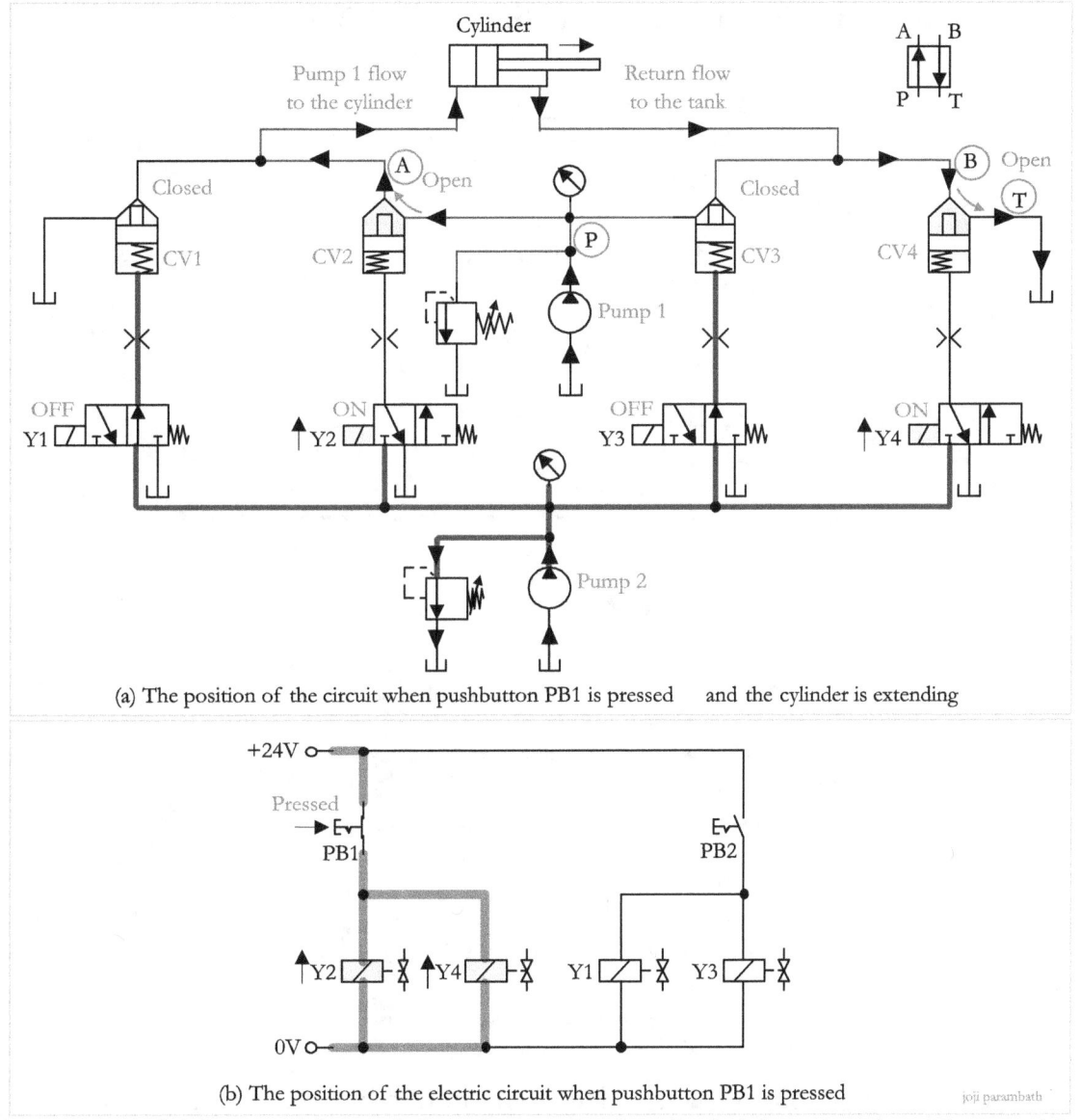

(a) The position of the circuit when pushbutton PB1 is pressed and the cylinder is extending

(b) The position of the electric circuit when pushbutton PB1 is pressed

Figure 4.10 | The position of the electro-hydraulic bridge circuit with cartridge valves when pushbutton PB1 is pressed (Example 4.6)

When pushbutton PB1 is pressed, solenoids Y2 and Y4 are energized.

The following sequence of actions takes place: (1) pilot ports of the cartridge valves CV2 and CV4 relieve pressure, (2) cartridge valves CV2 and CV4 open, (3) pump 1 flow is directed to the piston-side port of the cylinder, (4) the piston-rod-side of the cylinder is connected to the tank, and (5) the cylinder extends.

The Position of the Circuit When Pushbutton PB2 is Pressed

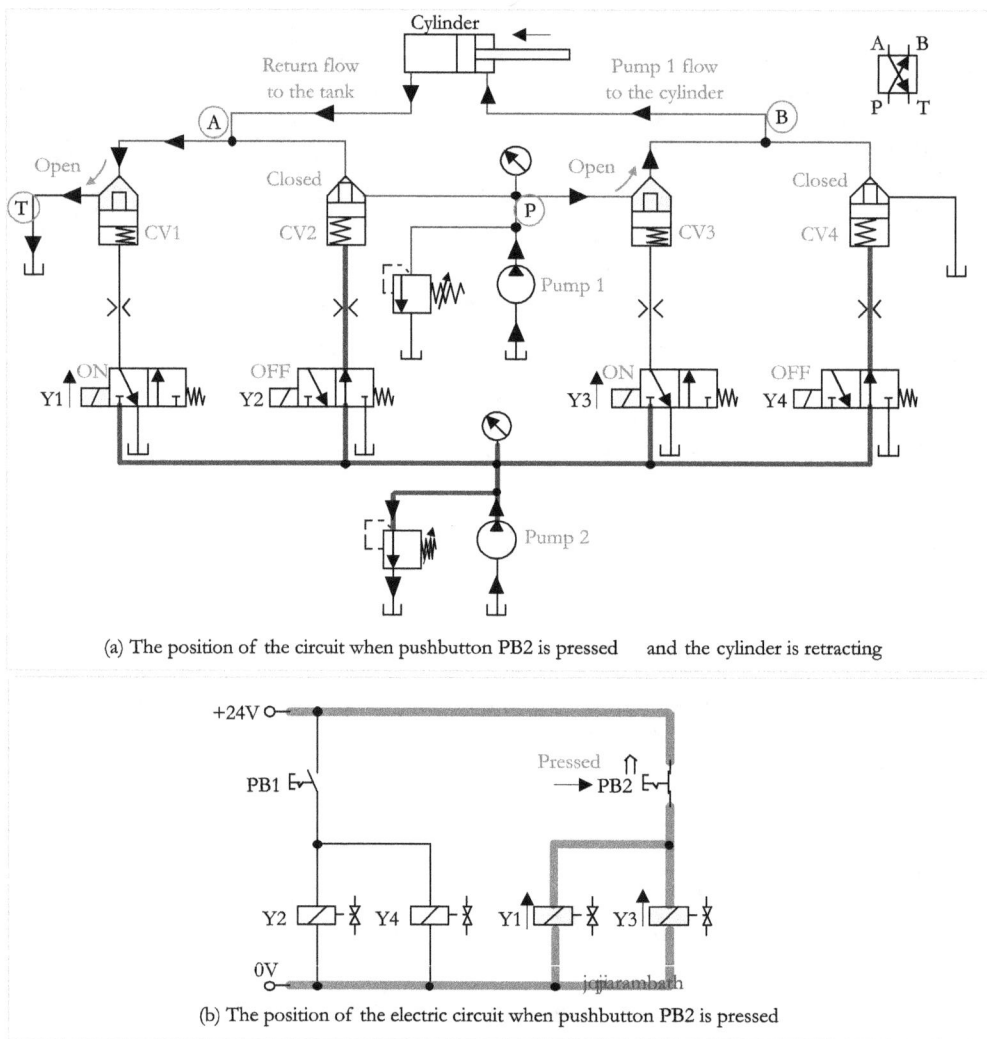

(a) The position of the circuit when pushbutton PB2 is pressed and the cylinder is retracting

(b) The position of the electric circuit when pushbutton PB2 is pressed

Figure 4.11 | The position of the electro-hydraulic bridge circuit with cartridge valves when pushbutton PB2 is pressed (Example 4.6)

The position of the electro-hydraulic circuit when pushbutton PB2 is pressed for the return motion of the double-acting cylinder is shown in Figure 4.11. When pushbutton PB2 is pressed, solenoids Y1 and Y3 are energized. The following sequence of actions takes place: (1) the pilot ports of the cartridge valves CV1 and CV3 relieve pressure, (2) cartridge valves CV1 and CV3 open, (3) pump flow is directed to the piston-rod-side port of the cylinder, (4) the piston-side of the cylinder is connected to the tank, and (5) the cylinder retracts.

Example 4.7 | Directional Control of a Double-acting Cylinder Using Cartridge Valves in Bridge Circuit Arrangement (Alternative Circuit)

Develop an electro-hydraulic circuit with four cartridge valves in a bridge circuit arrangement and an appropriate 4/3-way solenoid-operated DC valve for realizing the switching positions to control a double-acting cylinder's forward and return strokes.

Solution: Multiple positions of a circuit are presented in Figure 4.12.

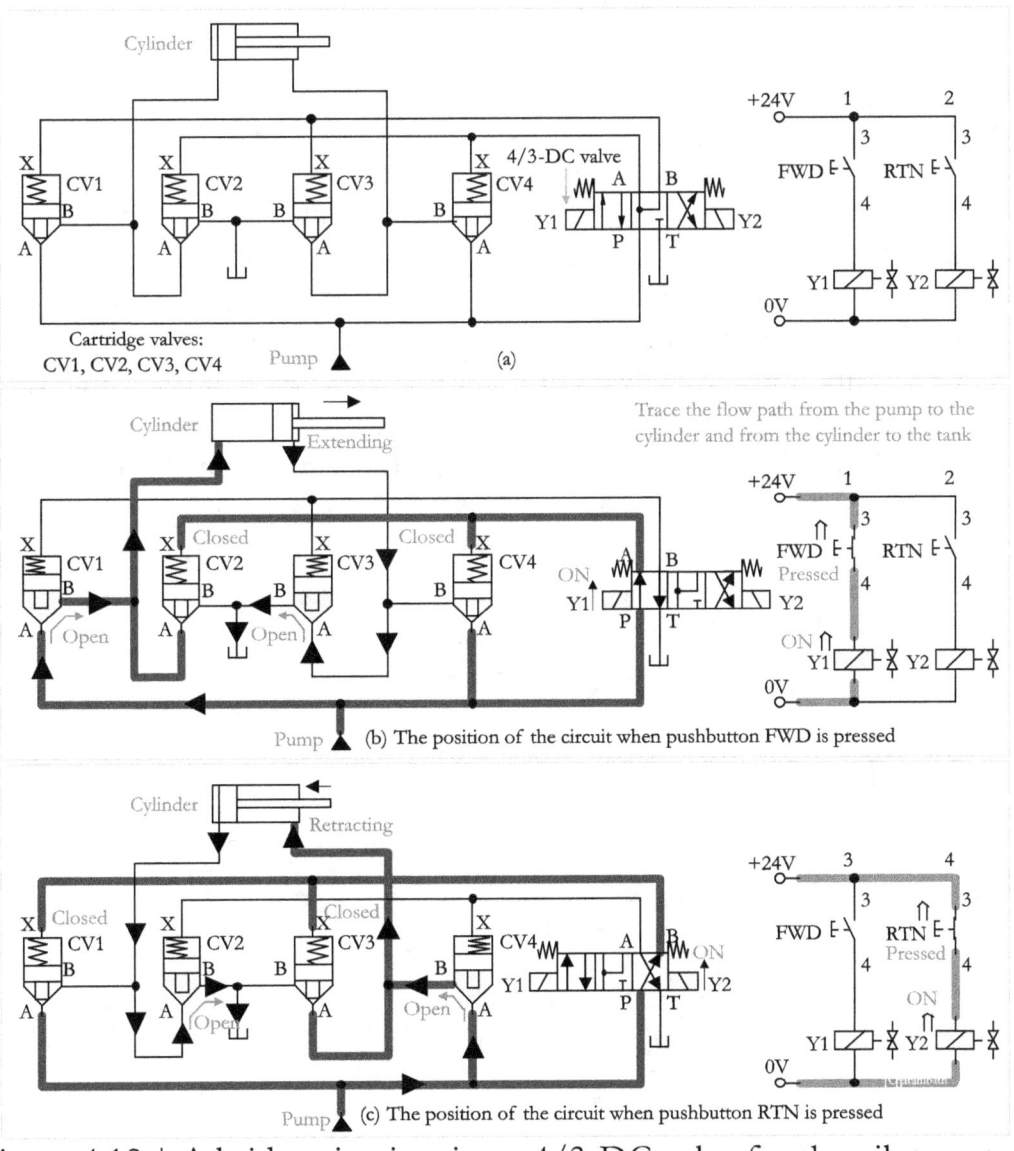

Figure 4.12 | A bridge circuit using a 4/3-DC valve for the pilot control

Chapter 5 | Control Covers and Circuits for Pressure Controls

Hydraulic pressure functions, such as pressure relief, unloading, and pressure reduction, can be achieved by combining a cartridge valve with appropriate control covers. A cartridge inserted in a pressure control valve typically includes a sleeve, a poppet with an area ratio of 1:1, and a closing spring. It is retained in the manifold cavity by a control cover. The control cover contains a manually adjustable pilot valve and piloting connections. Suitable orifices can be added to the pilot circuit to match application requirements.

Pressure can be set manually or electronically. Manual adjusters include a micrometer with or without a key lock and a standard square-end screw with a hexagonal locknut. Pressure can be set electronically through a proportional valve. Many standards specify the mounting interfaces for cartridge valves used in pressure-control applications. ISO 7368 specifies the position of the orientation pin for use with a main system relief valve. This feature ensures that no other valve function is installed where a system relief valve is required.

A Basic Cartridge Valve for the Pressure Relief Function

Figure 5.1 shows the cross-section and symbolic diagram of a cartridge-type pressure relief valve without an area differential. The valve can be integrated into or mounted onto the control cover. It also includes orifices and ports X, Z1, AP, and Y, as well as a knob for setting the pressure. Pressure can be set over a wide range, typically from 3 to 350 bar (43 to 5000 psi).

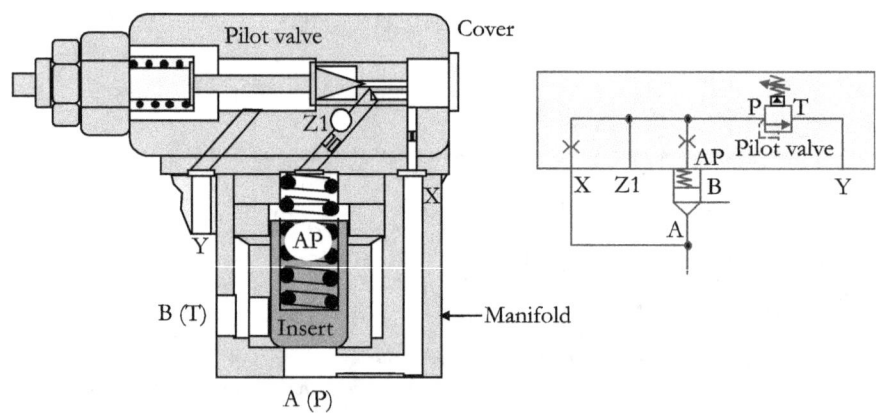

Figure 5.1 | A basic pressure relief valve in cartridge form

The orifices allow the fluid to flow through the port AP to the pilot area of the associated poppet and the pilot valve within the cover. When the pilot pressure reaches the pilot valve setting, the valve opens, relieving the pilot pressure in the spring chamber AP to the tank through port Y.

Other Variants of Pressure Relief Valves in Cartridge Form

Figure 5.2 shows a cartridge configuration with a fixed orifice in the poppet connecting port A to the pilot relief stage. This allows through-spool piloting for faster response. The pilot valve setting determines the pressure at port A.

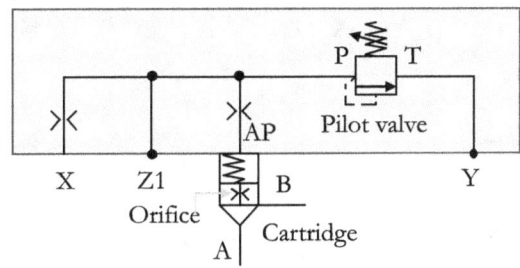

Figure 5.2 | A cartridge-type pressure relief valve with an orifice in the insert

Figure 5.3 shows a pressure-relief valve cartridge configuration with a single solenoid pilot valve. When the solenoid is de-energized, the cartridge valve's spring chamber is vented through the solenoid valve and port Y. When the solenoid is energized, pressure at port A is limited by the pilot valve's setting in the cover.

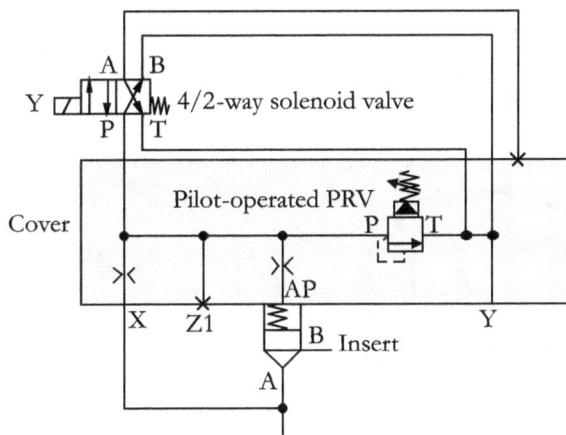

Figure 5.3 | A cartridge-type pressure relief valve with a single solenoid valve

Example 5.1 | Configure a cartridge valve system with a 1:1 ratio insert, a control cover with a pressure relief valve, and a single solenoid pilot valve for the following function. The system with a fixed-displacement pump must limit the pressure in the associated hydraulic line when the single solenoid valve is energized and vent the cartridge when the solenoid valve is de-energized. Develop the control circuit to realize the function.

Solution

Figure 5.4 shows the configuration of the cartridge valve with the 1:1 ratio insert, the control cover with the pressure relief valve, and the single solenoid pilot valve. The control cover includes ports X, Z1, AP, Y, P, T, B, and A, and the insert includes ports A and B.

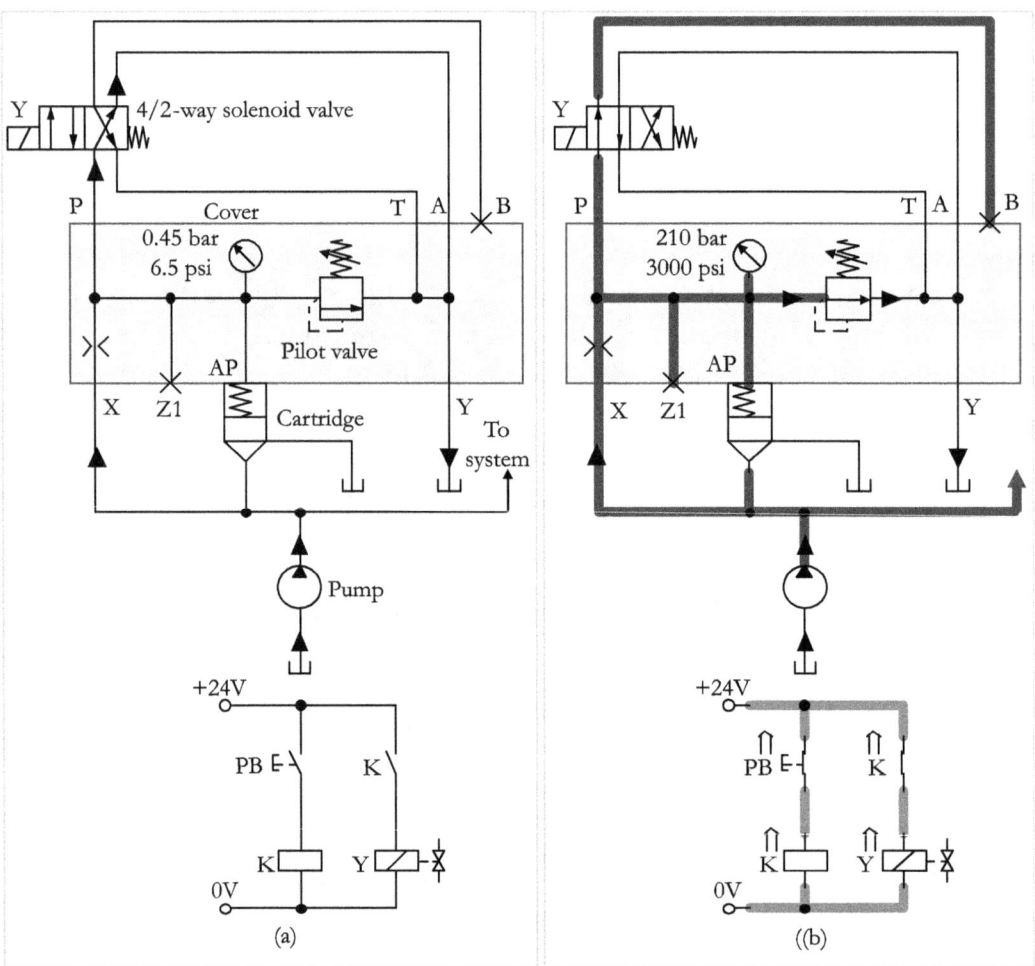

Figure 5.4 | A cartridge valve circuit for the pressure relief function

As shown in Figure 5.4, the cartridge is vented when the solenoid is de-energized. When the solenoid is energized, the pressure at port A is limited to the pressure-relief valve setting.

Example 5.2 | Control of a Hydraulic Cylinder Using a Cartridge Valve with a Pressure Relief Function

Develop a hydraulic circuit to control a double-acting cylinder using a cartridge valve with a 1:1 ratio insert and a control cover with a pressure relief valve.

Solution

Figure 5.5 shows multiple circuit positions for controlling a double-acting cylinder using a 4/2-way double solenoid valve with solenoids Y1 and Y2. A fixed-displacement pump supplies the required fluid for operation. The pressure can be set by using a cartridge-type pressure relief valve. The cartridge system consists of an insert, a control cover integrated with a pressure relief valve, and a 4/2-way single solenoid with solenoid coil Y for the pilot control of the cartridge.

Figure 5.5(a) shows the circuit's position when the pump is not energized.

Figure 5.5(b) shows the position of the circuit when the pump is switched on. The pump flow is bypassed to the associated tank. The pressure, if any, in the spring chamber (AP) of the cartridge insert is relieved to the tank through the 4/2-way single solenoid valve. Alternatively, the pump flow can be relieved through port A to port B of the cartridge valve, depending on the pressure conditions prevailing across the insert.

Figure 5.5(c) shows the position of the circuit when solenoid Y is energized. In this position, the spring chamber (AP) is pressurized as the flow paths to the tank through the 4/2-way valve and port A to port B of the cartridge valve are blocked. The pressure can develop to the setting of the pressure relief valve.

Figure 5.5(d) shows the position of the circuit when the solenoids Y and Y1 are energized. In this position, the cylinder extends. When the cylinder reaches its fully extended position, the pressure again rises to the pressure relief valve setting.

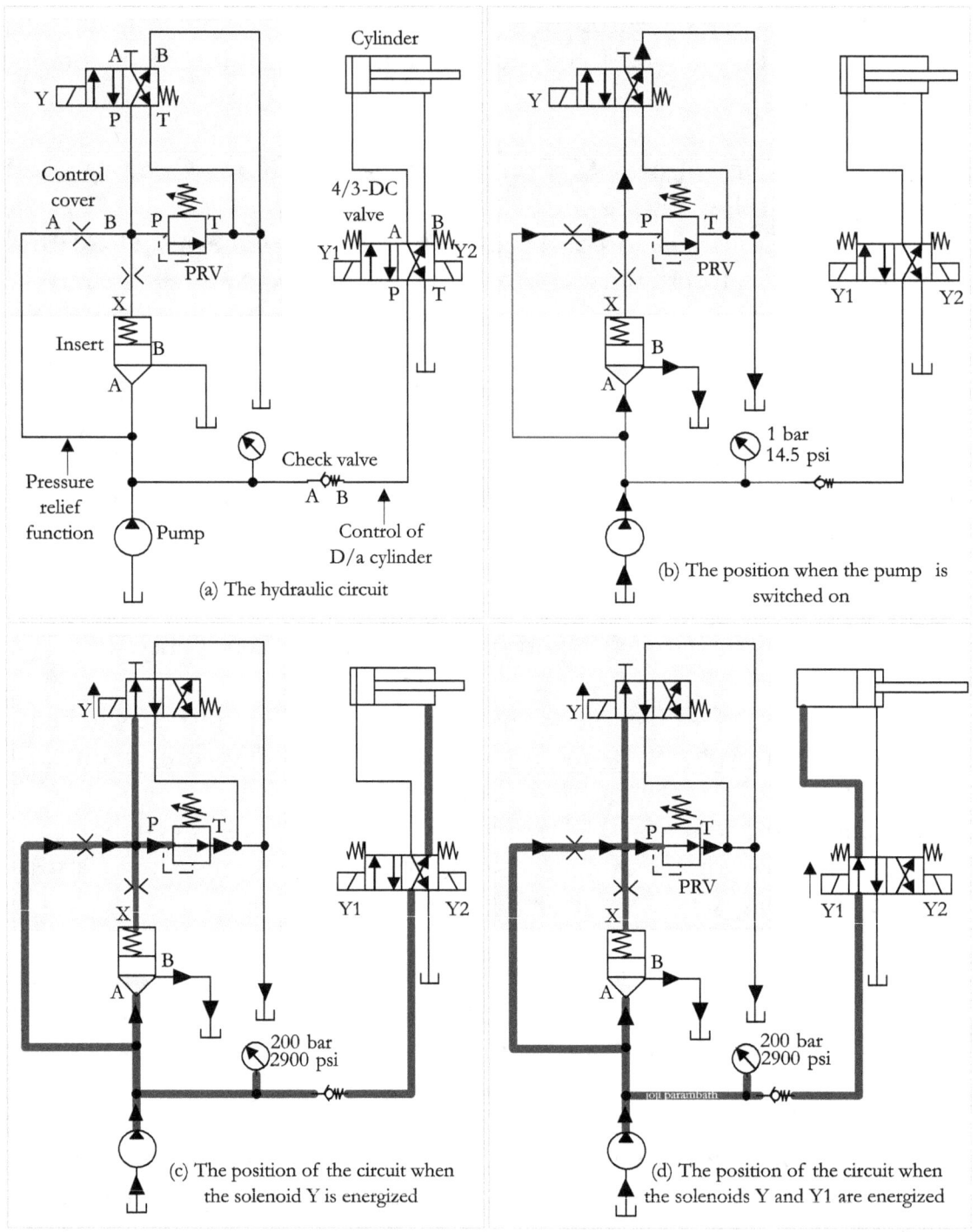

Figure 5.5 | Multiple positions of a circuit for the control of a hydraulic cylinder using a cartridge valve having a control cover with a pilot valve (Example 5.2)

A Cartridge Valve for the Unloading Function

A cartridge valve with an unloading function, as shown in Figure 5.6, can automatically load/unload a fixed delivery pump according to system demands. The cartridge insert is a sliding spool type with metering holes for a progressive opening. The control cover is integrated with an unloading valve. The pilot port of the unloading valve is accessible externally through port X.

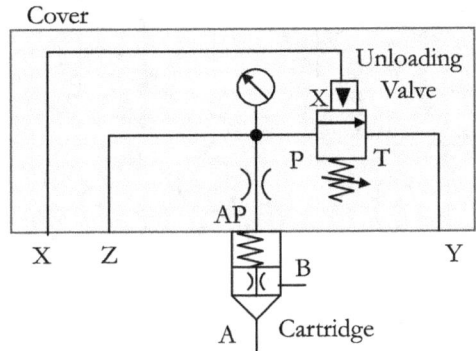

Figure 5.6 | A control cover for the unloading function

Typical examples of hydraulic systems that use unloading functions include an accumulator charging system for storing energy and a two-pump hi-lo system for driving a large cylinder.

Example 5.3 | Accumulator Charging System

A fixed-displacement hydraulic pump is used to load an accumulator for energy storage. The pump should automatically unload when the accumulator is fully charged to the required pressure. Develop a hydraulic accumulator load/unload system with a cartridge-type unloader.

Solution

Figure 5.7(a) shows a hydraulic circuit for automatically loading and unloading an accumulator using a cartridge-type unloader. The required pressure is set on the unloading valve. The fixed-displacement pump is connected to the accumulator through a check valve. The check valve allows the pump to charge and pressurize the accumulator and prevents reverse flow.

The high-pressure side downstream of the check valve is connected through port X to the pilot port of the unloading valve in the control cover.

The unloading valve opens when the accumulator pressure reaches the unloading valve setting. This causes the pressure to drop in the cartridge valve's spring chamber. Consequently, the main poppet of the cartridge valve lifts fully, opening the flow path from port A to port B, as shown in Figure 5.7(b).

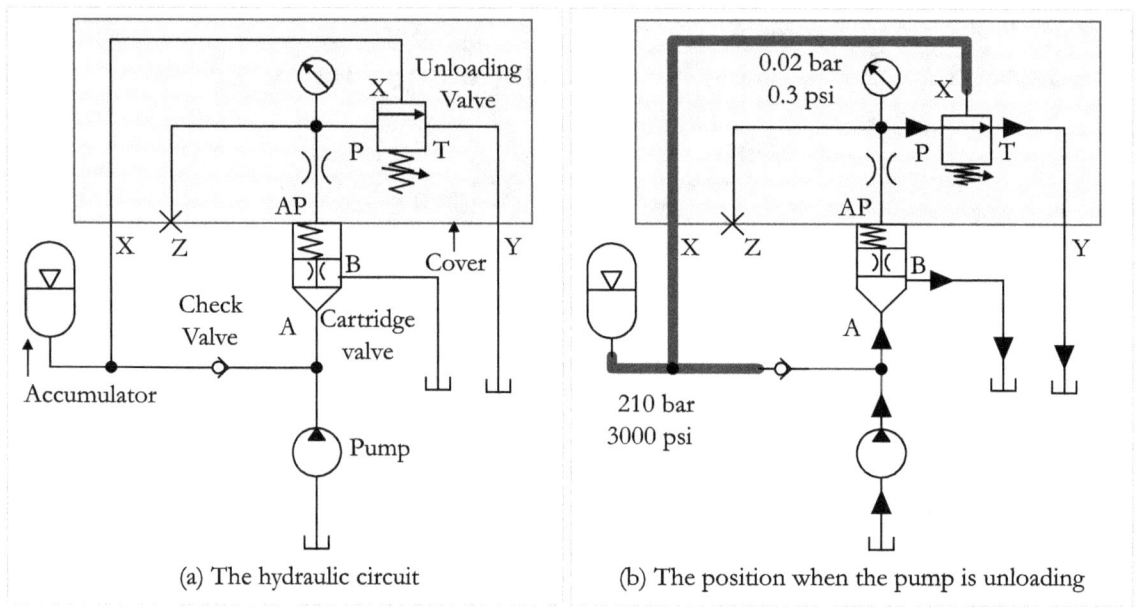

Figure 5.7 | Multiple positions of a circuit for an accumulator charging system

Two-pump 'hi-lo' Power Supply

Many hydraulic systems demand high flows at low pressures for the rapid motion of high-volume cylinders and then high pressures at low flows for feeding or clamping operations. The well-known 'hi-lo' circuit can realize this type of requirement.

A hi-lo circuit uses two hydraulic pumps. One pump, known as a volume pump (P1), is designed to provide a high flow at low pressure. The second pump, known as the pressure pump (P2), is designed to deliver high pressure at low flow. Initially, the combined outputs from pumps P1 and P2 give the maximum flow to the system at low pressure. When the load pressure reaches the unloading valve setting (e.g., 25 bar), the high-flow, low-pressure volume pump (P1) is bypassed through the unloading valve to the tank, allowing the low-flow, high-pressure pump (P2) to supply all system requirements.

Example 5.4 | Hi-lo Circuit)

A hi-lo hydraulic system is to be designed with two pumps (namely, a volume pump and a pressure pump) to provide high flow at low pressure initially and high pressure at low flow subsequently to control a large-volume hydraulic cylinder. The cylinder must extend rapidly with the high-flow, low-pressure fluid produced mainly by the volume pump until it reaches the work point. At this point, the cylinder must operate only with the high-pressure, low-flow fluid delivered by the pressure pump. Develop a cartridge valve control circuit to implement the scheme.

Solution

Figure 5.8 gives two critical positions of the hi-lo control circuit with the two pumps and the cartridge valve control circuit. Note that the hydraulic cylinder is not shown. Figure 5.8(a) shows the partial hydraulic circuit, and Figure 5.8(b) shows the position of the circuit when the volume pump is bypassed through the cartridge valve.

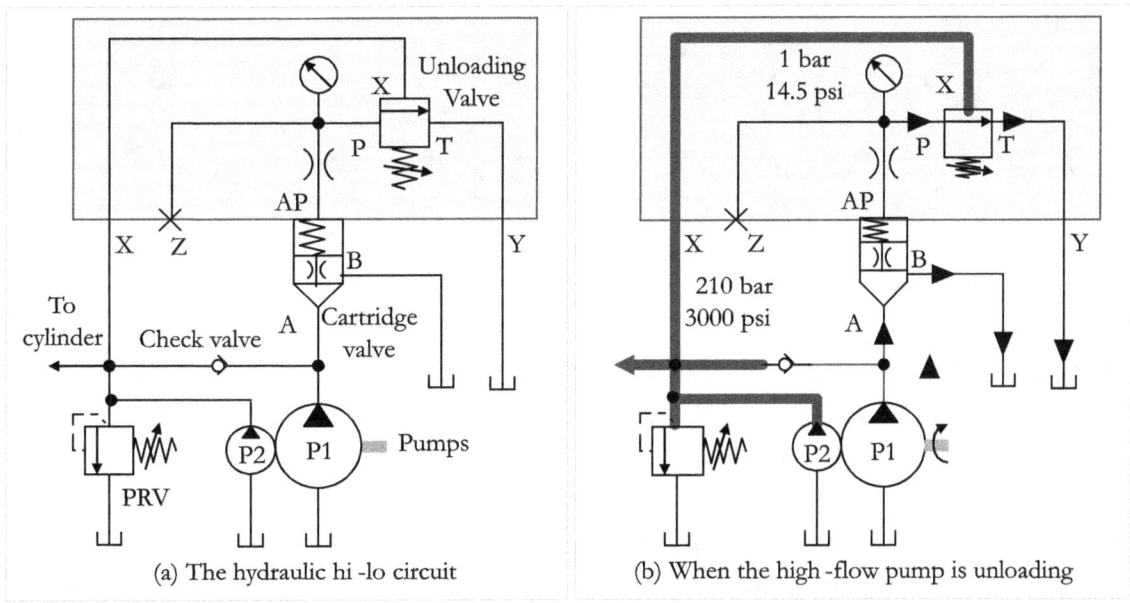

Figure 5.8 | Two positions of a double-pump hydraulic circuit (Example 5.4)

The check valve can isolate the low-pressure side from the high-pressure side. The pressure relief valve limits the maximum pressure (say, 100 bar) generated by the pressure pump.

A Cartridge Valve for Pressure-Reducing Function

As shown in Figure 5.9, the pressure-reducing cartridge valve consists of a spool-type insert and a control cover. The insert, with ports A (outlet) and B (inlet), includes a check valve. The insert closes the valve in the normal position. The cover includes a pressure-compensated flow-control valve and a pressure-relief valve with a manual adjuster. It also has ports X, Z1, AP, and Y. The valve is designed to provide a constant outlet pressure below the inlet pressure. Port B is connected to port X.

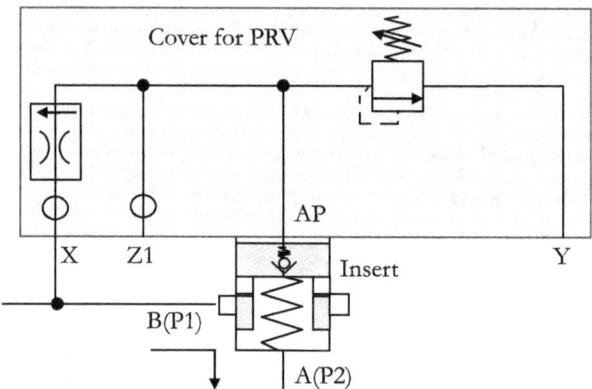

Figure 5.9 | A cartridge valve for the pressure-reducing function

The system pressure at port B is directed to area AP through the pressure-compensated flow-control valve in the cover. This valve maintains a constant flow to the pilot chamber AP, independent of the main flow from port B to port A, stabilizing pressure.

When the outlet pressure at port A is below the set pressure, the spool is held open by the spring force and the pressure differential acting on the insert. The fluid flows from port B to port A.

The pressure relief valve setting limits the pressure in the pilot chamber of the cartridge valve. Upon reaching the set pressure, the spool closes with only a slight increase in load pressure. The closed valve maintains this output (load) pressure. Load pressure transients are relieved through the check valve located in the insert. Port Z1 can be used to remotely control the reduced pressure. Venting of port Z1 will cause the outlet pressure at port A to drop to a minimum determined by the spring load in the insert. Port Z1 should be blocked when not required. Port Y is used to drain the pilot fluid.

Example 5.5 | Pressure Relief and Remote Electrical Venting

An electrohydraulic cartridge valve system must be designed to vent the hydraulic pump in its normal 4/2-way, solenoid-operated pilot-valve position, regardless of system demands. The pump should unload when the pilot valve is energized and the set pressure is reached. Develop an electro-hydraulic system to realize the functions.

Solution

Figure 5.10 shows the electrohydraulic cartridge valve system to realize the control functions. The system consists of an insert, a control cover with an unloading valve, and a 4/2-way solenoid-operated pilot valve.

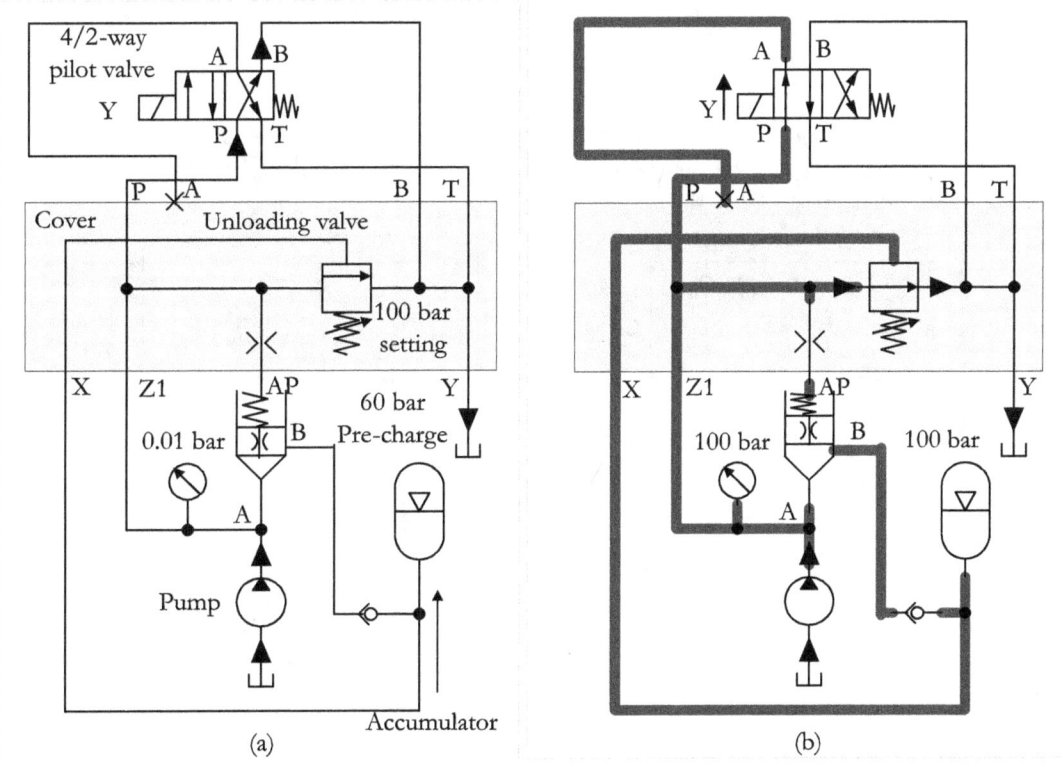

Figure 5.10 | An electro-hydraulic cartridge circuit for pressure relief and electrical venting

A pilot connection from port A to port Z1 is required to ensure the pressure relief function from port A to port B.

The pump flow at port A can be unloaded by applying pilot pressure to port X. Figure 5.10(a) shows that the system is vented when the solenoid is de-energized. Assume that a constant-displacement pump supplies the flow to charge an accumulator.

When the solenoid is energized, as shown in Figure 5.10(b), the unloader function operates through the pilot port X.

To initiate unloading, the pressure must equal the unloader pilot stage setting. The system pressurizes again when this pressure drops by approximately 20%.

Test Your Understanding

Study the two positions (a) and (b) of a cartridge valve pressure control circuit given in Figure 5.11.

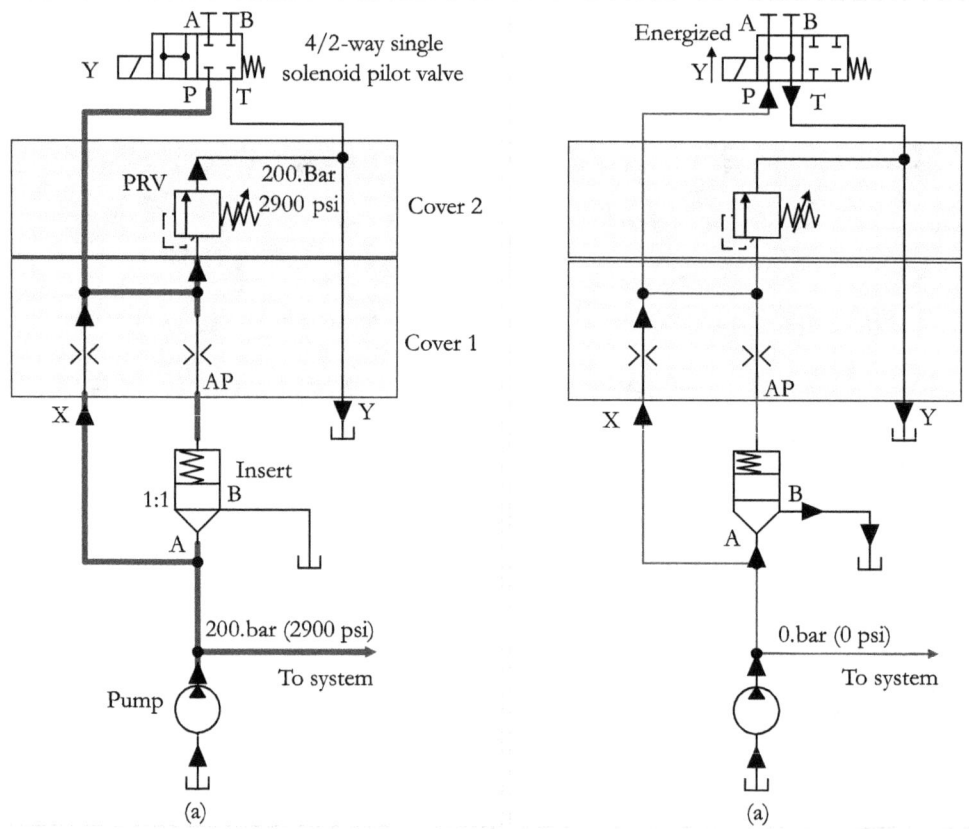

Figure 5.11 | Two positions of a pressure control cartridge valve circuit, normally closed for unloading

Chapter 6 | Control Covers and Circuits for Flow Controls

Figure 6.1 shows a cartridge valve with an insert and a cover for flow control. The insert has a metering notch for flow control. The control cover has a stem that limits the insert's stroke and flow. The flow can be adjusted by turning a knob, which moves the stem up or down to limit how far the poppet can open. When the pilot pressure is removed, the valve will open to the position set by the stem.

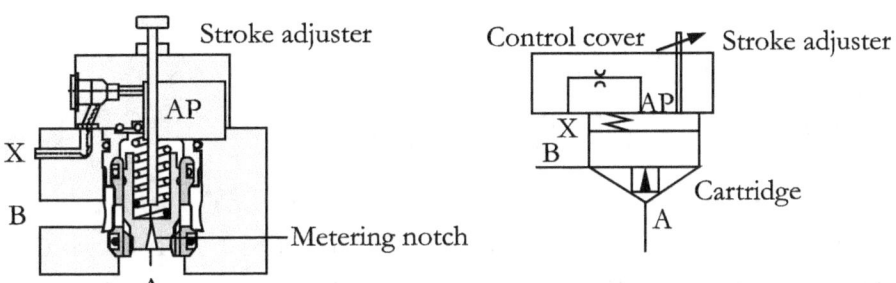

Figure 6.1 | A control cover with an adjustable stroke limiter

The adjustable stroke limiter includes a check or directional function, as shown in Figure 6.2. The directional function can be achieved using a basic configuration or a single-solenoid-controlled pilot valve on the control cover.

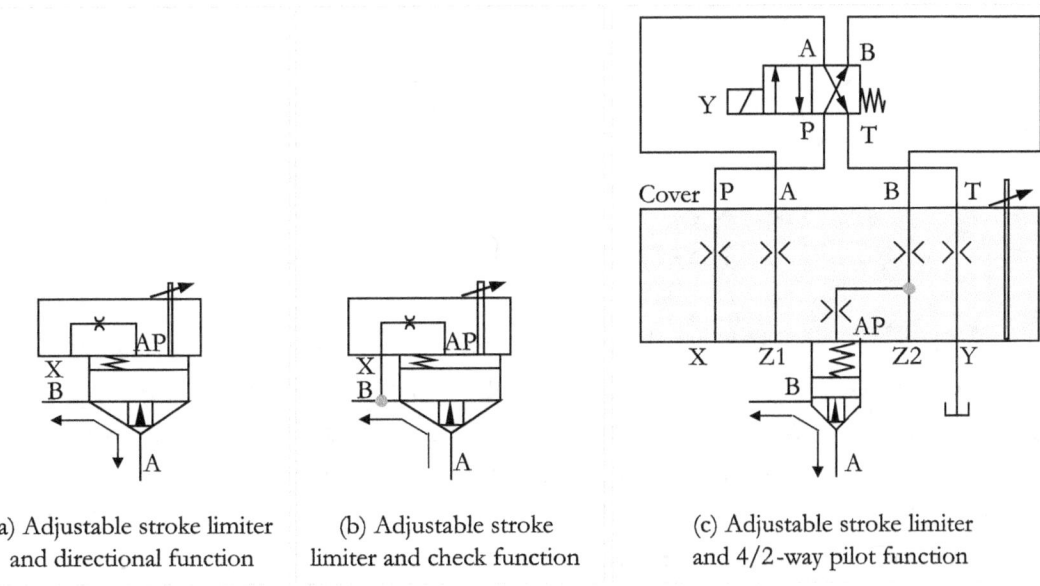

(a) Adjustable stroke limiter and directional function

(b) Adjustable stroke limiter and check function

(c) Adjustable stroke limiter and 4/2-way pilot function

Figure 6.2 | Variants of adjustable stroke limiters

Figure 6.2(a) shows an adjustable stroke limiter and directional function. The adjustable limiting of the insert poppet opening restricts flow in both directions (from port A to port B and from port B to port A). The external pilot signal is given through port X.

Figure 6.2(b) shows an adjustable stroke limiter and check function. Port X of the cover is connected to port B of the insert. The adjustable poppet lift limiter restricts flow from port A to port B, and the check function prevents flow from port B to port A.

Figure 6.2(c) shows an adjustable stroke limiter function and a pilot control through a 4/2-way single-solenoid valve.

Example 6.1 | Configure a cartridge valve system with an insert and a control cover with an adjustable stroke limiter for the check/flow control function. Develop the control circuit to realize the function.

Solution

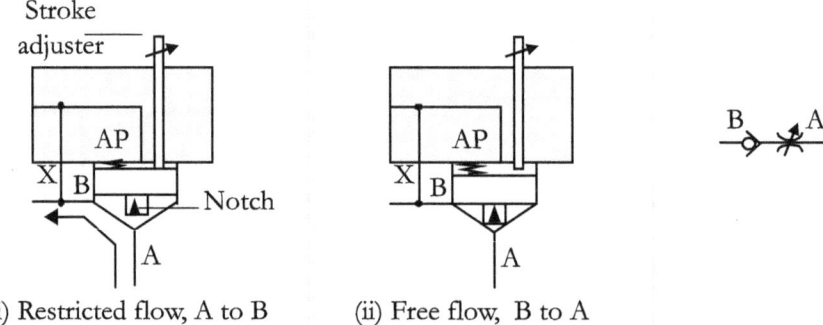

(i) Restricted flow, A to B   (ii) Free flow, B to A

Figure 6.3 | A cartridge valve configuration with a check/flow control valve

Figure 6.3 shows the cartridge valve configuration for the check/flow control function, which is self-explanatory. Because port B is connected to the spring chamber (AP) through port X, the cartridge valve configuration acts as a check valve, permitting flow from port A to port B but blocking flow from port B to port A. The metering notch on the insert provides flow control. The stroke adjuster limits the stroke of the insert and the flow. The flow can be adjusted by turning the adjuster knob, which moves the stem up or down to limit how far the poppet can open.

Chapter 7 | 3-way/4-way Spool-type Cartridge Valves

Previous chapters described 2-way poppet logic valves. Poppet logic valves are commonly used for high-flow directional switching by employing small, low-power pilot valves to control the sequence of operation. A single logic valve can be used for 2-way on/off switching, while multiple elements in a bridge arrangement can control 3-way or 4-way directional switching.

Cartridge-type 3-way 2-position, 4-way 2-position, and 4-way 3-position solenoid-operated spool valves are also available with flow rates up to 35 lpm (9 gpm) and pressures up to 350 bar (5000 psi). Spool-type cartridge valves can be used for directional switching. However, they are typically used to control flow or regulate pressure in modulating applications.

They offer a wide choice of flow paths and operating positions and can easily be integrated into manifold solutions to satisfy the most demanding application requirements.

3-way 2-position Spool-type Cartridge Valves

Figure 7.1 shows a 3-way 2-position, solenoid-operated, spool-type, screw-in cartridge valve.

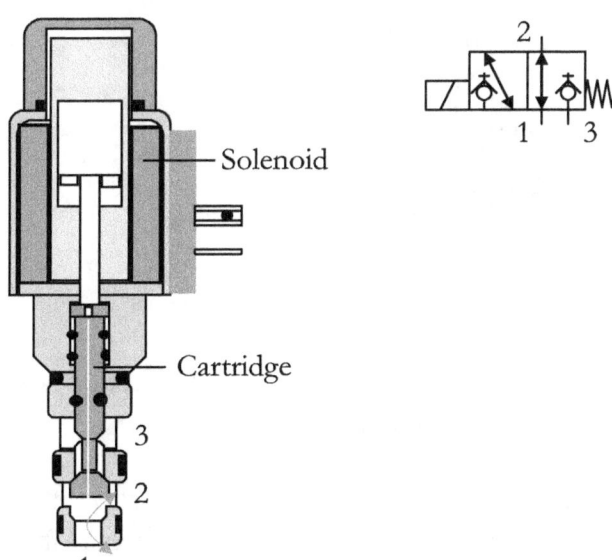

Figure 7.1 | A 3-way 2-position, solenoid-operated, spool-type, screw-in cartridge valve

In the normal position, when the solenoid is de-energized, the valve allows flow bidirectionally from port 1 to port 2 while blocking flow at port 3. When the solenoid is energized, the spool moves and opens flow bidirectionally from port 2 to port 3 while blocking flow at port 1. It may be noted that it has a spool-type valving mechanism rather than a poppet-type mechanism for port 2-to-port 3 switching.

4-way 2-position Spool-type Cartridge Valves

Figure 7.2 shows a 4-way 2-position, solenoid-operated, spool-type, screw-in cartridge valve.

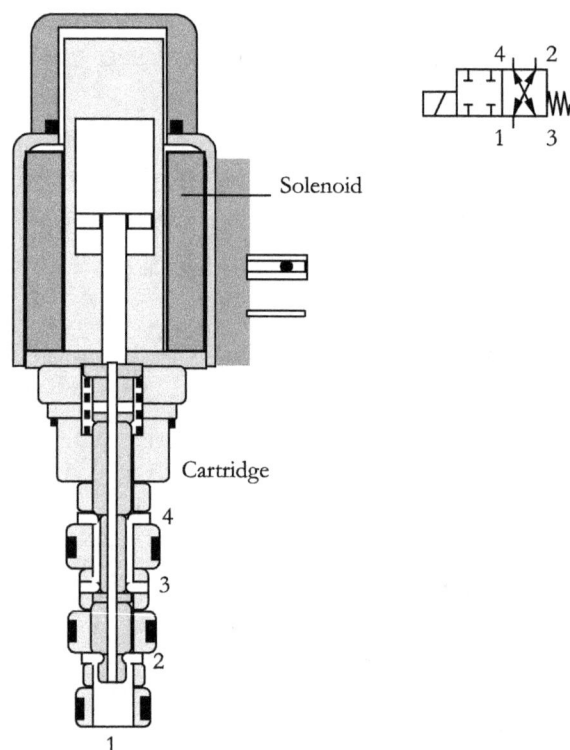

Figure 7.2 | A 4-way 2-position, solenoid-operated, spool-type, screw-in cartridge valve

In the normal position, when the solenoid is de-energized, the valve allows flow from port 1 to port 2 and from port 4 to port 3, bidirectionally.

When the solenoid is energized, the spool shifts, blocking all ports; it has a spool-type valving mechanism rather than a poppet-type.

An Overview of 3-way Spool-type Cartridge Valves

Figure 7.3 shows symbols of 3-way spool-type cartridge valves.

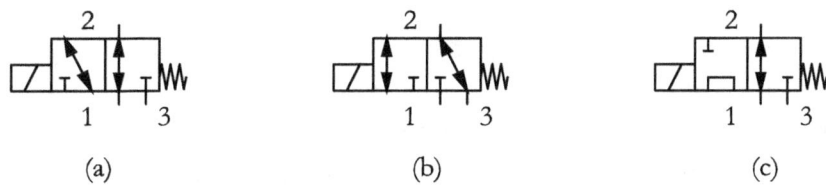

Figure 7.3 | Symbols of 3-way spool-type cartridge valves

An Overview of 4-way 2-position Spool-type Cartridge Valves

Figure 7.4 shows symbols of 4-way 2-position spool-type cartridge valves.

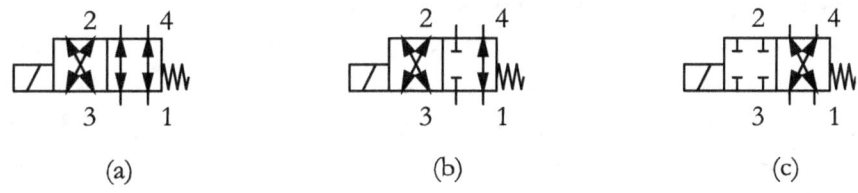

Figure 7.4 | Symbols of 4/2-way spool-type cartridge valves

An Overview of 4-way 3-position Spool-type Cartridge Valves

Figure 7.5 shows symbols of 4-way 3-position spool-type cartridge valves.

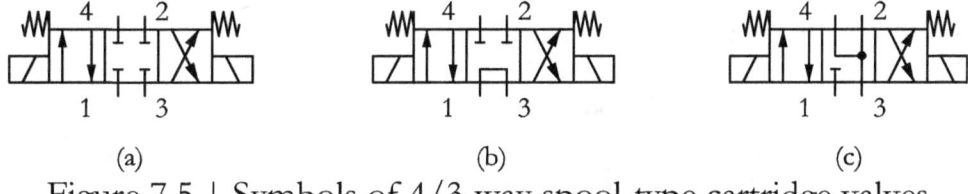

Figure 7.5 | Symbols of 4/3-way spool-type cartridge valves

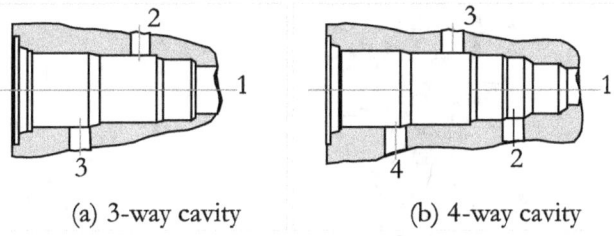

(a) 3-way cavity   (b) 4-way cavity

Figure 7.6 | Identification of cavities

Chapter 8 | Actively-controllable Cartridge Valves and Circuits

A cartridge valve logic assembly with only one control area in its spring chamber is regarded as a passive logic valve. In contrast, a logic assembly having a differential insert with two control areas is termed an active logic or dynamic valve. The dynamic insert extends above the manifold in an intermediate cover, creating an additional control area. The pilot pressure in the additional control area can keep the active logic valve open even when there is no pressure in port A or port B. The actively controlled logic assembly is designed to be compact, modular, and fast-acting.

2-way Cartridge valve, Actively Controllable

Figure 8.1 shows a 2-way, actively controllable cartridge valve and an equivalent symbolic representation. It consists of a control spool (cartridge), an intermediate cover, and a control cover. The valve has two main ports, A and B, and two pilot ports, X and Y, on the intermediate cover. The pilot ports are used to control the dynamic insert remotely.

The spring chamber in the intermediate cover has a differential spool. The spool has areas A1, A2, and A4 in the opening direction and area A5 in the closing direction. The effective force acting on the spool determines its position and motion. The pilot pressure in the control area (through Y) can keep the active logic open even with no pressure on ports A or B.

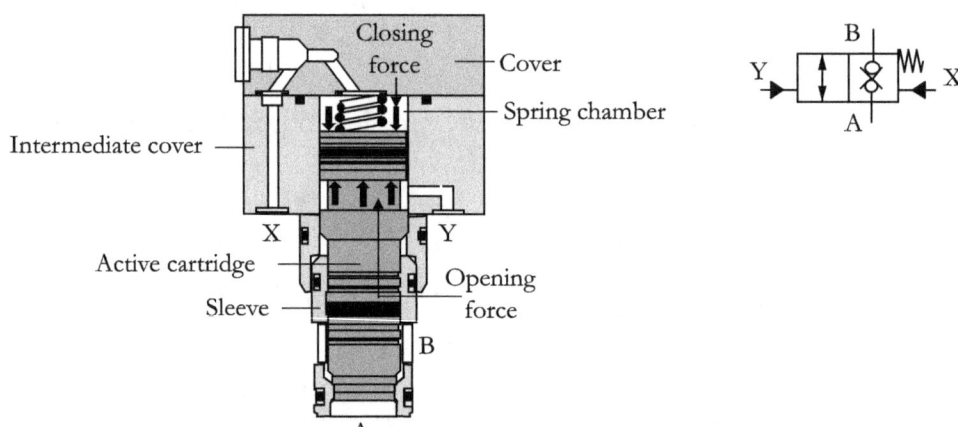

Figure 8.1 | 2-way cartridge valve, actively controllable

This control cover connects to the pilot control valves and/or other hydraulic elements, thereby integrating the functions. All pilot and poppet seals create a tight fit at all ports to prevent leakage in either direction.

Type of Standard Inserts, Active Cartridge Valves

Active cartridge valves use various types of cones and sleeves. Three basic types are shown in Figure 8.2.

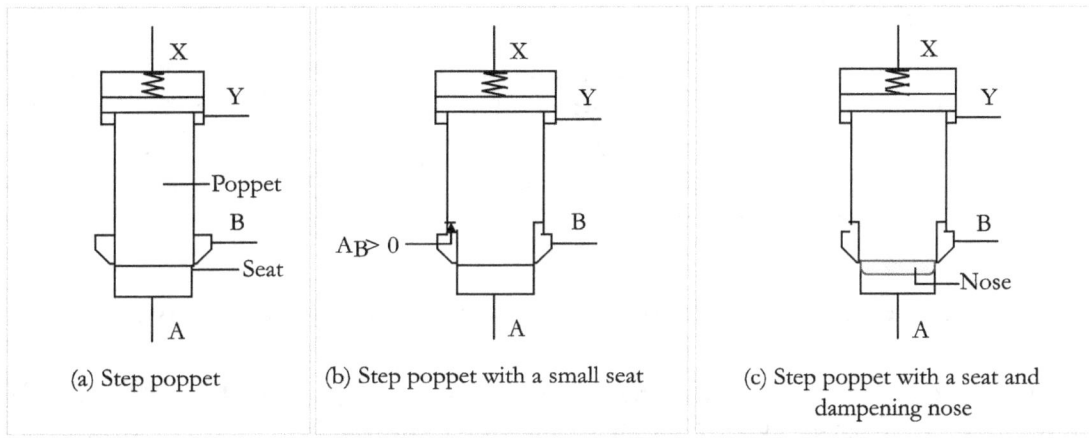

Figure 8.2 | Types of inserts of active cartridge valves

Step Poppet

See Figure 8.2(a). The cartridge valve's poppet and sleeves are designed without an effective area for port B ($A_B = 0$). Additionally, the poppet has no dampening nose. This type of active cartridge valve can be used for all directional and check functions where a larger flow area and lower pressure losses are desired.

With a non-actuated control ($Y = 0$), the flow of this cartridge with a non-differential-area poppet can only flow from port A to port B.

Step Poppet with a Small Seat

See Figure 8.2(b). The cartridge valve's poppet and sleeves are designed with an effective area for port B ($A_B > 0$). The poppet has no dampening nose. This type of active cartridge valve can be used for all directional, check, and flow control functions. With a non-actuated control ($Y = 0$), the flow of this cartridge with a differential-area poppet can be selected (Port A to Port B or Port B to Port A).

Step Poppet with a Small Seat and a Dampening Nose

See Figure 8.2(c). The cartridge valve's poppet and sleeves are designed with an effective area for port B ($A_B > 0$). Additionally, the poppet has a dampening nose. This type of active cartridge valve can be used to avoid additional pressure peaks in tank circuits or to adjust flow-control valves more accurately.

Symbol, Active Cartridge Valve

Figure 8.3 shows a typical circuit with a 2-way cartridge valve.

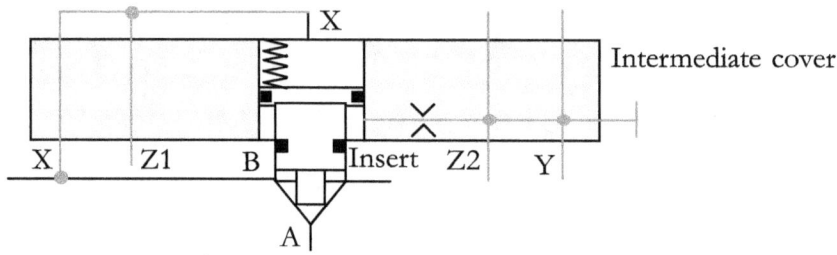

Figure 8.3 | The symbol of a 2-way active cartridge valve

The active logic valve can be directly installed in a standard installation bore according to ISO 7368. Thus, it is also suitable for retrofitting an existing passive logic valve that must be leakage-free inside, requires position monitoring, or achieves faster closing times.

Control Covers, Active Cartridge Valves

Many control covers for active cartridge valves are designed to realize many control functions. Figure 8.4 shows two types of control covers.

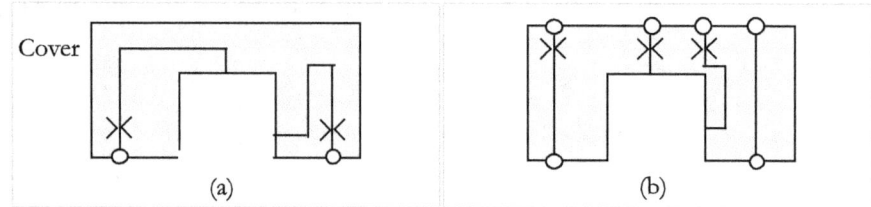

Figure 8.4 | Types of control covers of active cartridge valves

Changes in the metering-in and metering-out flow through the pilot lines can influence a cartridge valve's function and switching velocity. This could be realized by changing mounting orifices as required.

Example 8.1

Draw the symbol of an active cartridge valve with an intermediate control cover suitable for a 4-way solenoid-operated pilot valve to obtain a normally open function.

Solution

Figure 8.5 shows an active cartridge valve with a basic intermediate cover. Ports P, A, B, and T interface with the 4-way solenoid-operated pilot valve.

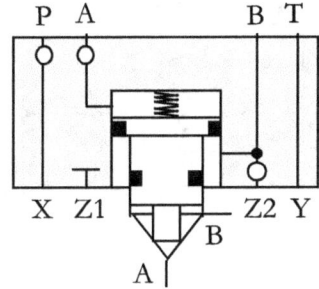

Figure 8.5 | An active cartridge valve with a basic intermediate cover

Example 8.2

Draw the symbol of an active cartridge valve with an intermediate control cover suitable for a 4-way solenoid-operated pilot valve to obtain a normally open function and a stroke limiter.

Solution

Figure 8.6 shows an active cartridge valve with a basic, intermediate cover and a stroke limiter. Ports P, A, B, and T interface with the 4-way solenoid-operated pilot valve.

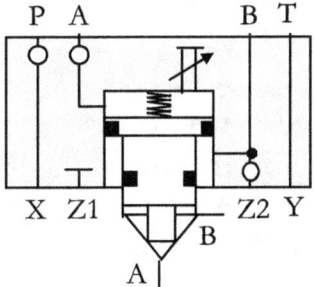

Figure 8.6 | An active cartridge valve with a basic, intermediate cover and a stroke limiter

Configurations, Active Cartridge Valves

Active cartridge valves can be configured using inserts, intermediate and control covers, and solenoid-operated directional control valves. Figure 8.7 shows typical configurations.

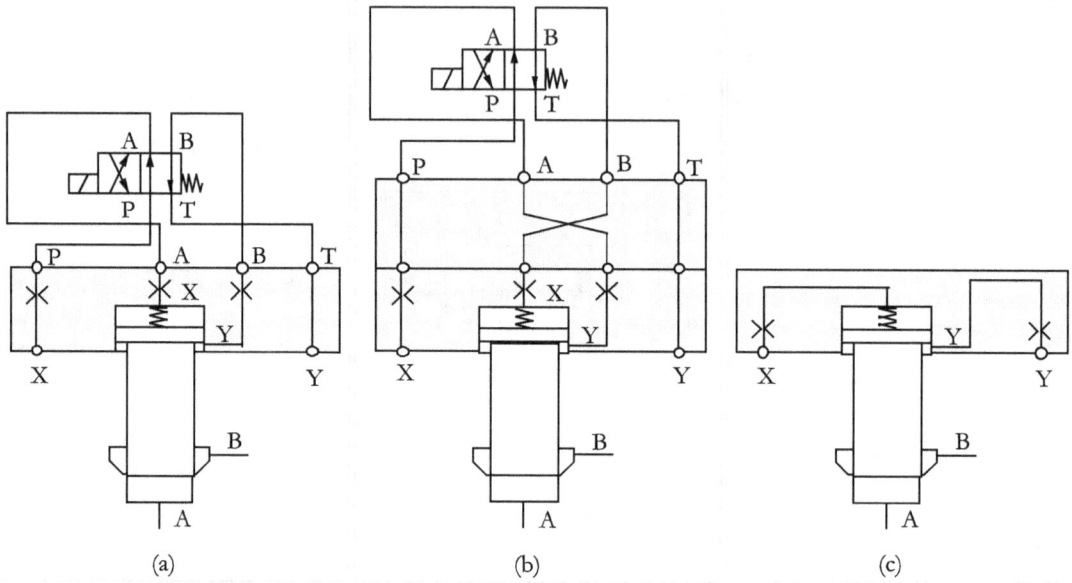

Figure 8.7 | Configurations of active cartridge valves

Figure 8.7(a) shows a cartridge valve configuration with an active poppet, an intermediate cover, and a 4-way solenoid-operated pilot valve. In the de-energized state of the pilot valve, the closing area of the cartridge valve's spring chamber (X side) is energized through port P to port A flow path of the pilot valve, and the pressure from the opening area (Y side) of the spring chamber is relieved through port Y. Therefore, the cartridge valve remains tightly closed. In the energized state of the pilot valve, the cartridge valve opens, allowing flow from port A to port B or from port B to port A.

Figure 8.7(b) shows a cartridge valve configuration with an active poppet, an intermediate cover, a reversing control cover, and a 4-way solenoid-operated pilot valve. In the pilot valve's de-energized state, the spring chamber's opening area (Y side) is energized via the port P-to-port B flow path, and the pressure in the closing area of the spring chamber (X side) is relieved. Therefore, the cartridge valve remains open, allowing flow from port A to port B or from port B to port A. In the pilot valve's energized state, the cartridge valve remains tightly closed.

Figure 8.7(c) shows a self-explanatory cartridge valve configuration with pilot ports X and Y operated with external pilot signals.

Features of Active Cartridge Valves
Many important features of active cartridge valves are listed below.

- The opening and closing of active cartridge valves are independent of port A or port B pressure.
- They are highly reliable and offer a great degree of control and repeatability
- Cartridges of active cartridge valves can be fully interchangeable with conventional cartridges.
- They are designed to provide a low hysteresis and fast response.

Applications of Active Cartridge Valves
Active cartridge valves have been developed for applications that require fast opening and closing times and to ensure that the insert/poppet is seated positively.

An active cartridge valve is pilot-operated to cut off the main hydraulic supply to a clamp circuit. It is also used to control press closing. Active Poppet provides the required protection for clamp circuits.

Active cartridge valves can be used in various applications, including presses, injection molding machines, lifting equipment, and accumulator systems.

Safety, Active Cartridge Valves
Manufacturers typically set, test, and seal active cartridge valves. The settings should not be altered to ensure safety.

Example 8.3 | Control of an Accumulator Using an Active Cartridge Valve System
An active cartridge valve is used to control the discharge of fluid from a hydraulic accumulator to a single-acting cylinder. A fixed-displacement pump supplies the necessary fluid to the system. A modular-type throttle check valve controls the cartridge's opening speed. A metering orifice also controls the closing speed. Develop a control circuit.

Solution

Figure 8.8 shows a circuit controlling a hydraulic accumulator using an active cartridge valve. A constant-displacement pump supplies fluid to the system. The cartridge valve closes and opens the main flow path in response to pilot signals at its closing (X) and opening (Y) pilot ports. An orifice provided in the path of the pilot signal controls the speed of the cartridge valve's closing, and a throttle check valve controls the cartridge valve's opening speed.

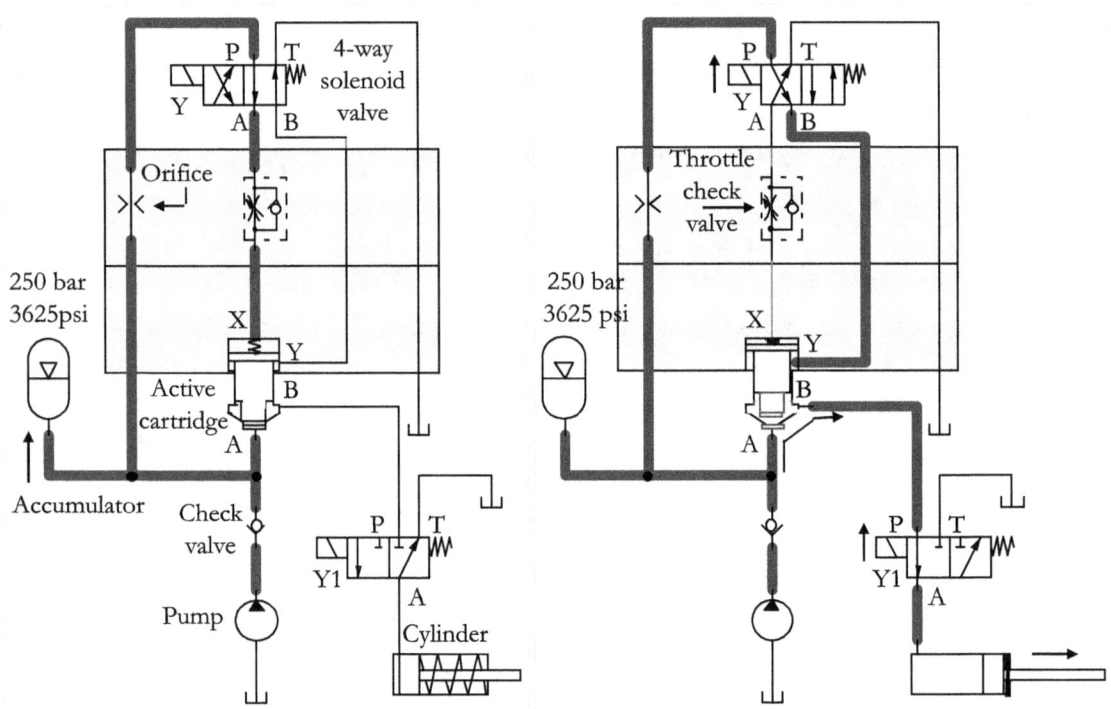

(a) The position when the solenoid is de-energized   (b) The position when the solenoid is energized

Figure 8.8 | Two positions of the circuit for the control of the accumulator using an active cartridge valve system

Figure 8.8(a) shows the position of the circuit when the solenoid of the 4-way pilot valve is de-energized. In this position, the pilot signal is directed to closing port X of the spring chamber through the orifice, and opening port Y of the spring chamber is relieved. Therefore, the cartridge valve is tightly closed. Figure 8.8(b) shows the position of the circuit when the solenoid of the 4-way pilot valve is energized. In this position, the pilot signal is directed to open port Y of the spring chamber, and port Y of the spring chamber is relieved through the throttle valve. Therefore, the cartridge valve remains open.

Chapter 9 | Proportional Cartridge Valves and Circuits

A basic cartridge valve system consists of an insert installed in the cavity of a manifold, with appropriate flow passages and a control cover. The insert has several metering notches (orifices) to control flow.

Proportional Flow Control Cartridge Valves

A proportional flow control valve can be constructed like a switching 2-way, 3-way, or 4-way cartridge valve. The insert can be controlled by a proportional solenoid, which, in turn, is driven by a current signal from an electronic controller. The required flow rate can be set using an input device such as a potentiometer or a joystick. When current flows through the solenoid, the insert moves to open the control notches and proportionally increases the cross-sectional area of the flow path. The control produces a flow output through the valve proportional to the input current signal.

There are various cartridge-style flow control valves, including in-line, 2-way, 3-way (priority), and 4-way. A pressure-compensated flow control valve, with the help of a pressure regulator, can provide a regulated flow proportional to the current input, regardless of load or system pressure. Figure 9.1 shows symbols of the following proportional cartridge-style flow control valves: (a) in-line pressure-compensated, (b) in-line priority pressure-compensated, (c) 2-way throttle, normally-closed, (d) 2-way throttle, normally-open, (e) 3-way throttle, and (f) 4-way throttle.

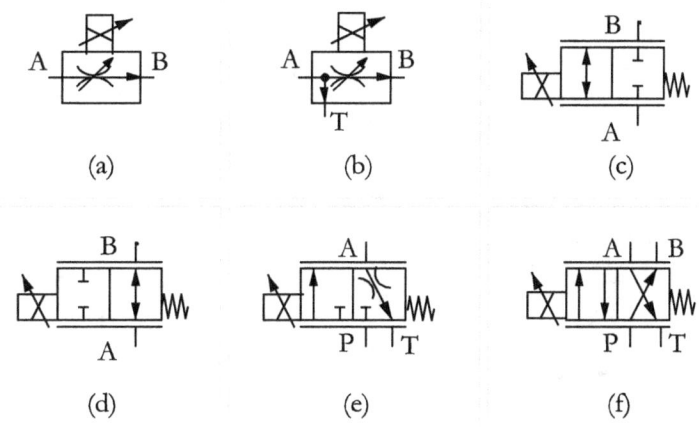

Figure 9.1 | Symbolic representations of proportional flow control valves

Proportional In-line Pressure-compensated Cartridge-style Flow Control Valve

Figure 9.2 shows a proportional in-line (restrictive) 2-way pressure-compensated screw-in cartridge-style flow control valve. It can be either normally open or normally closed.

When current flows through the solenoid, the insert moves to open control notches against a precision biasing spring, increasing the cross-sectional area proportionally. This results in a flow output through the valve proportional to the input current signal. The proportional valve continuously regulates the flow cross-section in response to a command input signal.

A pressure-compensator spool modulates the flow, typically at a ΔP of 6.9 bar (100 psi), providing the valve with a constant, regulated flow regardless of load or system pressure. The flow will increase with current for the normally closed valve and decrease with current for the normally open valve.

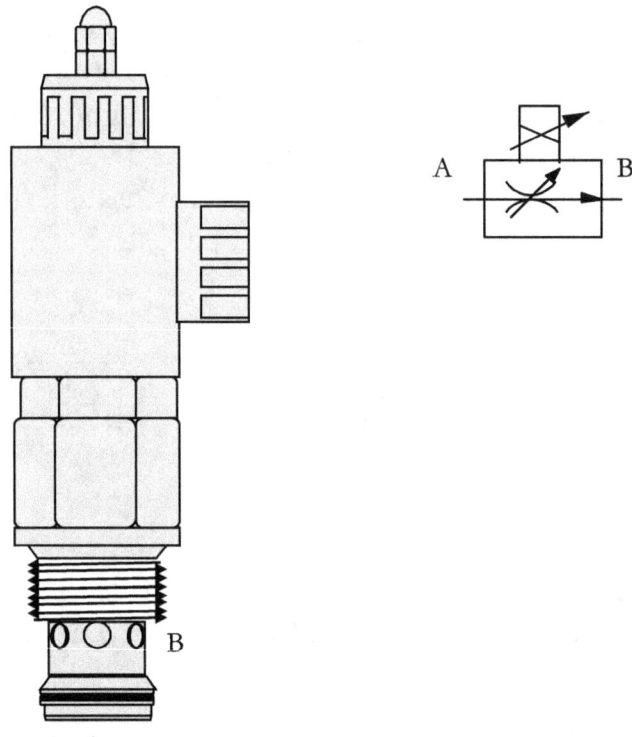

Figure 9.2 | A proportional in-line (restrictive) 2-way pressure-compensated screw-in cartridge-style flow control valve

Electro-hydraulic Proportional Priority Pressure-compensated Cartridge-style Flow Control Valve

Figure 9.3 shows a proportional-priority pressure-compensated screw-in cartridge-style flow-control valve. When current flows through the solenoid, the insert moves to open control notches against a precision biasing spring, increasing the cross-sectional area proportionally. This results in a flow output through the valve proportional to the input current signal. A pressure-compensator spool modulates the flow, typically at a ΔP of 6.9 bar (100 psi), providing the valve with a constant, regulated flow regardless of load or system pressure. After the priority flow is satisfied, the excess flow is diverted to a secondary circuit or the tank.

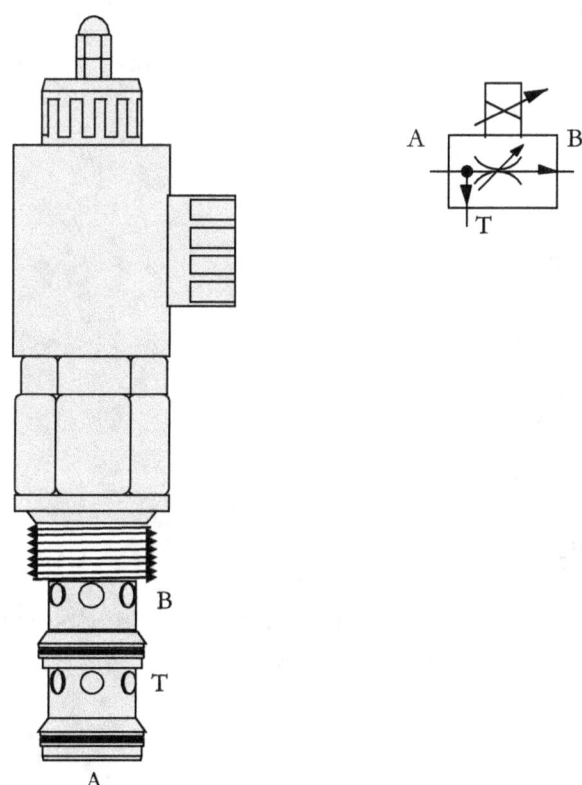

Figure 9.3 | A proportional in-line (restrictive) pressure-compensated cartridge-style flow control valve

A three-way pressure compensator ensures a stable pressure drop. The compensator can be modified as a two-way valve by closing the output channel. Electronics can be used to control the valve.

Proportional In-line, Non-compensated Cartridge-type Flow Control Valve

Figure 9.4 shows a proportional, in-line, non-compensated screw-in cartridge-type flow control valve. When current flows through the solenoid, the insert moves to open control notches against a precision biasing spring, increasing the cross-sectional area proportionally. This results in a flow output through the valve proportional to the input current signal.

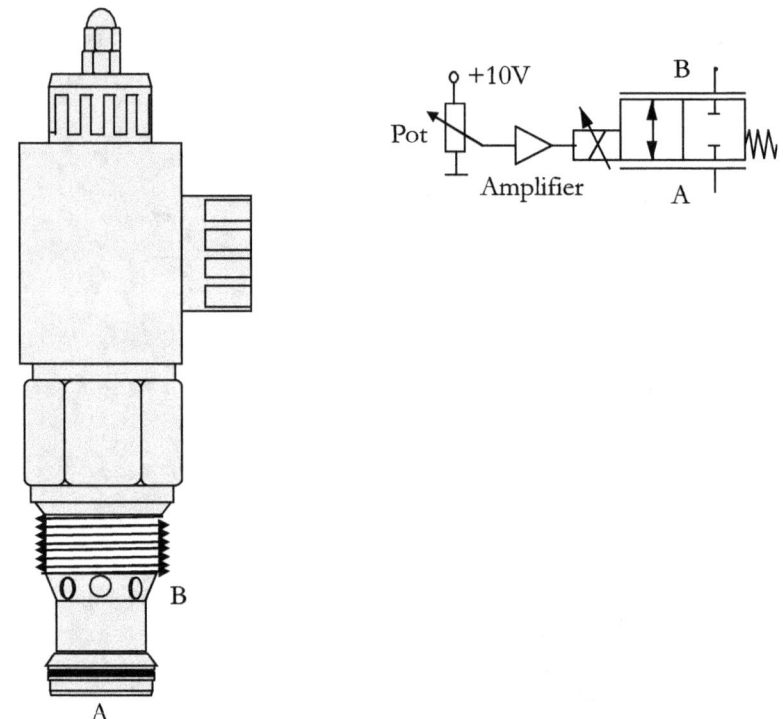

Figure 9.4 | A proportional in-line, non-compensated screw-in cartridge-type flow control valve

The proportional flow control valve is made up of a spool that is moved by the force of the proportional solenoid against the spring's restraining force. An amplifier controls the solenoid. The amplifier receives a command input signal from a potentiometer (pot), typically ranging from 0 to 10 volts. The amplifier processes the command input signal and generates a current signal proportional to the command input signal. As the current flows through the solenoid, it creates a magnetic force proportional to the current. The spool of the flow control valve moves in response to the magnetic force generated and produces a flow proportional to the command input signal.

Proportional 3/2-way Throttle Cartridge

Figure 9.5 shows a proportional 3/2-way throttle cartridge. It is designed on the sliding-spool principle. In the initial position (de-energized), port A is closed, and port B is connected to port T with the full flow rating. In control mode, the flow from port P to port A varied proportionally to the control current. Models are available for general use or special applications with or without compensators.

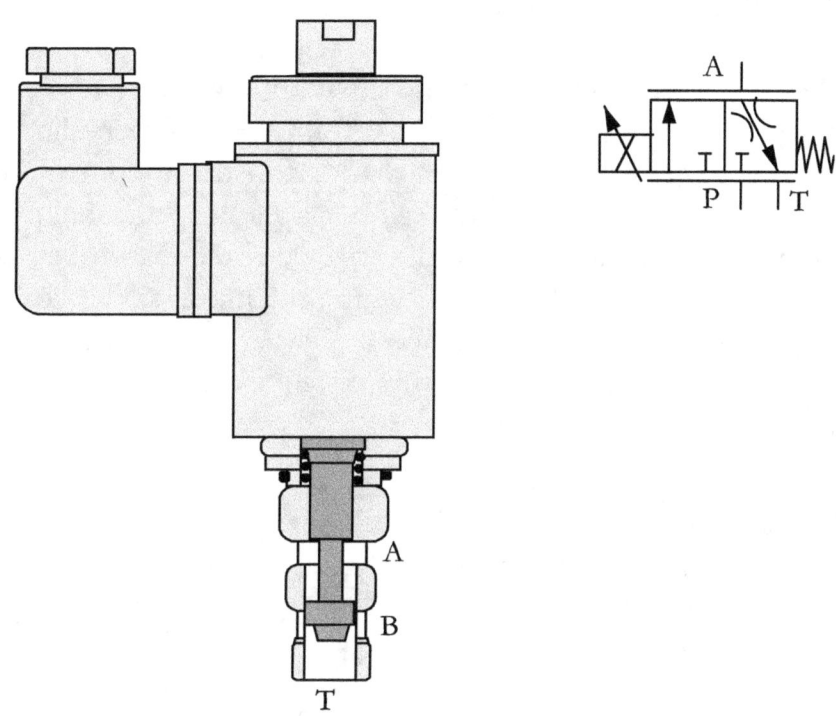

Figure 9.5 | Proportional 3/2-way throttle cartridge

Proportional 4/2-way Throttle Cartridge

Figure 9.6 shows the symbol of a self-explanatory 4-/2-way throttle cartridge.

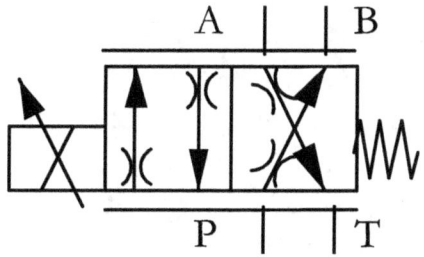

Figure 9.6 | Proportional 3/2-way throttle cartridge

Proportional Pressure Control Cartridge Valves

Proportional pressure control cartridge valves are components used in hydraulic systems to control fluid pressure. The control function can be pressure relief, unloading, or pressure reduction. They are designed to precisely regulate the pressure in a hydraulic system in response to an electrical signal. Proportional cartridge systems for pressure control primarily rely on the proportional pilot valve integrated into a control cover.

Proportional Pressure Relief Function, Cartridge Version

Figure 9.7 shows a cartridge configuration of an electro-hydraulic proportional pressure relief function. It consists of a cartridge insert, a control cover with a basic pressure relief valve, and a control cover with a proportional pressure relief valve with a current-controlled solenoid. An electronic controller controls the proportional pressure relief valve. The command input signal for the desired pressure setting can be provided remotely via an input device, such as a potentiometer or joystick. A proportional pilot valve can infinitely vary pressures over a wide range below this setting.

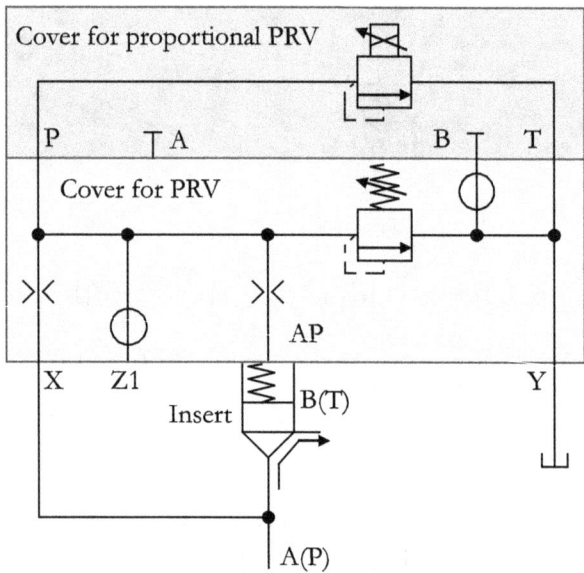

Figure 9.7 | A cartridge configuration of a proportional pressure relief function

The control covers for the proportional pilot PRVs and conventional PRVs should be selected according to the size and pressure range used. Typical maximum pressures for conventional PRV covers are 125, 250, and 350 bar

(1813, 3625, and 5000 psi). The pressure ranges for the pilot valve can be 5-40, 6-100, 7-160, 7.5-250, and 7.5-350 bar (72-580, 87-1450, 102-2320, 109-3625, and 109-5075 psi) for 16 mm size cartridges.

Proportional Pressure Reducing Function, Cartridge Version

As shown in Figure 9.8, the electro-hydraulic proportional pressure-reducing cartridge valve consists of a spool-type insert and control covers. The insert, with ports A (inlet) and B (outlet), includes a check valve. The insert closes the valve in the normal position. A cover includes a pressure-compensated flow-control valve and a pressure-relief valve with a manual adjuster. It also has ports X, Z1, AP, and Y. Another cover comprises a proportional pressure-reducing valve with a current-controlled solenoid. The interfacing ports of the covers are P, A, B, and T.

An electronic controller controls the proportional pressure-reducing valve. The command input signal for the desired pressure setting can be provided remotely via an input device, such as a potentiometer or joystick. Reduced pressure is proportional to the applied current. Increasing the current to the pilot increases the reduced pressure at port AP. The valve is designed to provide a constant outlet pressure below the inlet pressure. Port B is connected to port X.

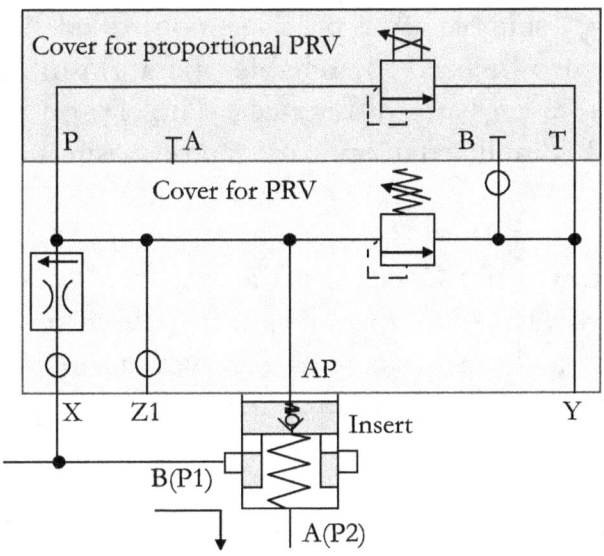

Figure 9.8 | A cartridge configuration of a proportional pressure-reducing function

The system pressure at port B is directed to area AP through the pressure-compensated flow-control valve in the cover. This valve maintains a constant flow to the pilot chamber AP, independent of the main flow from port B to port A, stabilizing pressure.

When the outlet pressure at port A is below the set pressure, the spool is held open by the spring force and the pressure differential acting on the insert. The fluid flows from port B to port A.

The PRV sets the maximum system pressure. The electronic pressure setting of the proportional pressure relief valve limits the pressure in the pilot chamber AP of the cartridge valve. When the set pressure is reached, the spool closes with enough load (output) pressure. The closed valve maintains this reduced load pressure. Load pressure transients are relieved through the check valve located in the insert.

Port Z1 can be used to remotely control the reduced pressure. Venting of port Z1 will cause the outlet pressure at port A to drop to a minimum determined by the spring load in the insert. Port Z1 should be blocked when not required. Port Y is used to drain the pilot fluid.

These cartridges are suitable for precise, controlled lifting and lowering movements and can also be used for reliable operation in mobile and industrial applications under large pressure differences. The slip-on coils can be replaced without opening the hydraulic envelope and can be positioned at any angle within 360°.

Salient Points, Proportional Cartridge Valves:
Proportional cartridge valves are appropriate for use in mobile hydraulics, such as excavators and agricultural machinery, as well as in industrial automation, including presses and plastic machinery. Key features of proportional cartridge valves include:
- compact design
- infinitely variable, smooth, and jerk-free control
- high precision and repeatability
- quick response time
- seamless integration with sensors and PLCs

Chapter 10 | Constructional Features of Integrated Manifolds

An integrated manifold combines multiple hydraulic components into a single unit. For example, it combines cartridge valves, pressure regulators, filters, accumulators, and other components into a single block. It is a distribution block that internally connects multiple fluid lines to components. It can route fluid flow, control its direction, and regulate pressure. It can also combine additional functionalities into the manifold itself. This integration of components into a single manifold simplifies the design, reduces space requirements, and improves efficiency. Figure 10.1 gives a graphic representation of the integrated manifold system.

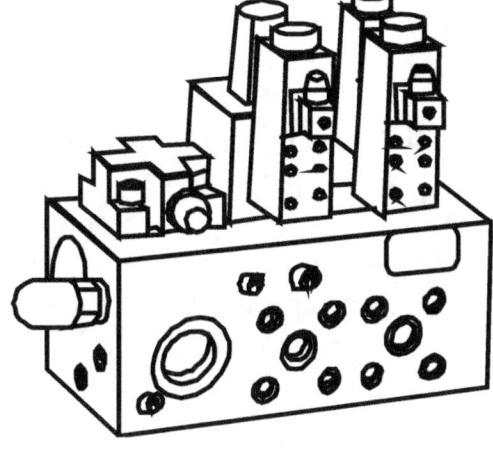

Figure 10.1 | An integrated manifold system

A hydraulic system may have many spool-type control valves and other components with complex circuitry. Other components may include filters, accumulators, and accessories. An alternative to the spool-type valves is the poppet-type cartridge valves. Combining these valves and components into an integrated manifold is often the most economical way of organizing a hydraulic system. This method makes the system compact, fast-acting, reliable, and leak-proof. This method is particularly useful when the flow rate exceeds 150 lpm (40 gpm), and the working pressure exceeds 210 bar (3000 psi).

Integrated manifolds are commonly used across industries such as aerospace, automotive, manufacturing, mobile systems, and industrial automation, where space, weight, and efficiency are critical.

Steps for the Construction of an Integrated Manifold

Some necessary steps for the design and construction of an integrated manifold system are listed in the bulleted lines given below:

Identify the Requirements: Determine the conditions, including the expected performances, load-induced pressures, and system safety factors. Specify the integrated manifold's functions, fluid types, pressure ratings, temperature ranges, and size constraints. The manufacturer and the customer may agree on performance requirements and test specifications before production.

Analysis: It is essential to determine how to interconnect the working ports to the cartridge cavities.

Sizing: This step involves selecting the correct material sizes for the cartridges, ports, and interconnections. The interconnections should be sized to match the flow requirements without excessive pressure drops.

Create a design model: This step creates a detailed 2D or 3D manifold model that incorporates all necessary components and features. A customer- or product-specific assembly drawing can be prepared for the manifold. Ensure design meets industry standards and regulations.

Machining and Assembly: In this stage, the manifold is machined, the cartridges are prepared to the required specifications, and the valve is assembled. Use machining processes like milling, turning, drilling, and grinding. Choose materials for the manifold based on fluid compatibility, strength, corrosion resistance, and manufacturability. Manifolds are machined from aluminum, ductile iron, or steel. CNC machining or additive manufacturing centers prepare them to maintain the highest quality standards. The manifolds should undergo extensive deburring and cleaning operations. Ensure tight tolerances and smooth surface finishes for optimal performance. All aluminum manifolds are anodized for cleanliness, have added surface hardening, and have enhanced corrosion resistance. Ductile iron or steel manifolds can be zinc-plated.

Assemble the manifold following the design instructions, using appropriate fasteners and seals to prevent leaks.

Testing: The final product is tested for leaks, pressure resistance, flow rates, and component functionality (e.g., valves, regulators, and filters) on automated computerized test stands to ensure quality and compliance with specifications. The test results are documented, and any necessary adjustments or corrections are made accordingly.

Advantages of Integrated Manifolds

Integrating multiple cartridge valves and other hydraulic components into manifolds significantly benefits industrial and mobile hydraulic systems.

In conventional spool-type valves, the timing of the opening or closing of all four ports occurs simultaneously, which can cause shocks. However, a proper cover can be selected in a cartridge valve to control the poppet's opening and closing times and travel. This feature enables the control of the acceleration and speed of the associated actuator.

The shocks can be reduced by properly sizing the orifices in the valve cover and timing the opening and closing of individual cartridges. A cartridge valve can be sized to handle only the flow required through its associated ports.

An integrated manifold is usually custom-made to meet the performance and installation needs of the specific machine. It consolidates the hydraulic control system into a compact and neat assembly, minimizing external connections, saving space and weight, and reducing external leakage, installation time, and system maintenance. Eliminating hoses, tubes, and fittings in manifold systems dramatically reduces installation costs and system maintenance.

Applications of Integrated Manifolds

Common applications of the integrated manifold include:
- Aerial Work Platforms
- Construction Equipment
- Farm Machinery
- Lift Trucks
- Refuse Management Equipment
- Road Maintenance Equipment
- Utility Service Equipment

Chapter 11 | Typical Characteristics and Specifications of Cartridge Valves

Cartridge valves for directional and check functions are essentially hydraulically piloted check valves. They are designed to handle nominal flows of up to 2800 lpm (740 gpm) per cartridge. These valves offer control options, including single- or multiple-pilot arrangements, flow restrictors, and solenoid-controlled pilot-operated directional control.

Cartridge valves can be arranged as single-function, multi-function, and integrated cartridges. The flow between ports A and B can be restricted, while the internal pilot orifice between port A and the spring chamber AP helps control the valve's operation. Poppets can be combined with different springs to provide different cracking pressures.

Pressure Ratings of Cartridge Valves, Typical

The maximum permissible pressure at ports A and B is typically 350 bar (5000 psi). The maximum pressure rating of all other ports, X, Z1, Z2, AP, Y, P, T, A, and B, is generally 350 bar (5000 psi). The minimum pressure is 0.3 to 6 bar (4.4 to 87 psi), depending on the poppet–spring combination.

Nominal Sizes and Flow Ratings of Cartridge Valves, Typical

A chart for flow ratings is given in Table 11.1 for guidance.

Table 11.1 | Typical nominal sizes and flow ratings of cartridge valves

Nominal size to ISO 7368	Nominal flow at ΔP of 1 bar (14.5 psi)		Nominal flow at ΔP of 5 bar (72 psi)	
	lpm	gpm	lpm	gpm
NG 16	90	24	230	60
NG 25	210	55	550	145
NG 32	425	112	900	238
NG 40	650	172	1200	317
NG 50	900	238	1700	449
NG 63	1200	330	2800	740

Functions of Cartridge Covers

The cartridge valves have basic check, pilot-operated check, directional, and flow control functions.

Typical Specifications of Cartridge Valve Inserts and Covers for Check, Directional, and Flow Restrictor Functions

Table 11.2 provides a chart of typical specifications for cartridge valve inserts and covers for check, directional, and flow restrictor functions (for guidance purposes).

Table 11.2 | Typical specifications of cartridge valve inserts and covers for check, directional, and flow restrictor functions

Parameter	Example
Operating pressure	420 bar (6000 psi)
Nominal size (NG)	16, 25, 32, 40, 50 63
Function (Ratio)	1:1, 1:1.05, 1:1.6, 1:1.2
Cracking pressure	0.31 to 6 bar (4.5 to 87 psi)
Cover function	Basic check, pilot-operated check, standard directional, pilot shuttle, stroke adjuster,
Internal leakage	0.10 cc/min at 420 bar (2 drops/min at 6000 psi)
Standard check bias spring	1 bar (14.5 psi)
Fluid operating temperature	-20° to +120°C (-4° to 248°F)
Fluid Compatibility	Mineral-based or synthetic
Viscosity	7.4 to 420 cSt (50 to 2000 SUS)
Filtration	21/19/16 or cleaner (per ISO 4406) Use with a filter rated ß10 ≥ 200
Cartridge Material	Steel with hardened work surfaces Zinc-plated exposed surfaces
Seal material	Buna N or Viton® O-rings PTFE backup rings

Typical Performance Data of Cartridge Inserts for Check, Directional, and Flow Restrictor Functions

Figure 11.1 shows a typical characteristic of the flow rate versus pressure drop through a cartridge insert.

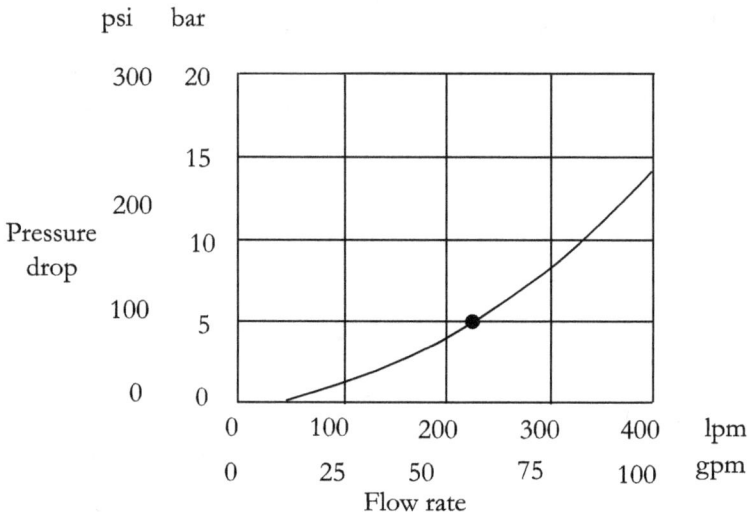

Figure 11.1 | A characteristic of flow rate/pressure drop through an insert

Leakage Characteristics of Cartridge Valves

Figure 11.2 shows a typical leakage characteristic of a cartridge valve at 36 cSt.

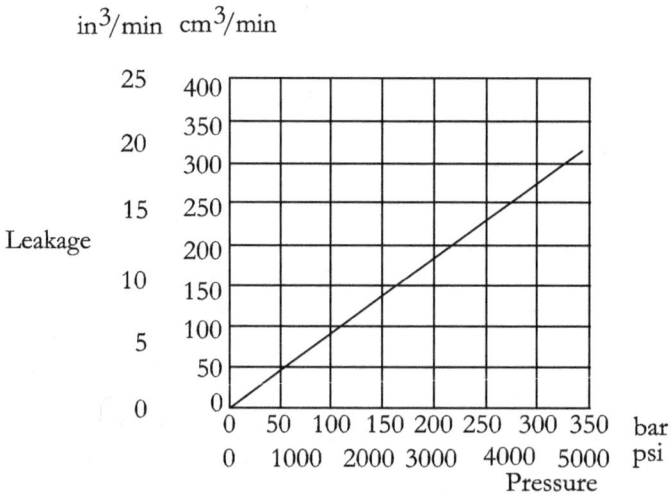

Figure 11.2 | Leakage characteristic of cartridge valves

Typical Specifications of Cartridge Version of Pressure Relief Function

Table 11.3 provides a chart of typical cartridge specifications for the pressure relief function (for guidance purposes).

Table 11.3 | Typical specifications of the cartridge type of pressure relief function

Parameter	Examples
Operating pressure	350 bar (5000 psi)
Nominal size (NG)	16, 25, 32, 40, 50 63
Nominal flow	13.3 lpm (3.5 gpm)
Function	1:1 ratio
Cracking pressure (for flow direction A to B)	0.31 to 2.35 bar (4.5 to 34 psi)
Standard check bias spring	1 bar (14.5 psi)
Cover function	Pressure relief
Adjuster mechanism	Micrometer adjuster with/without keylock Wrench adjustment with a hexagonal locknut
Adjustable pressure range	3 to 125 bar (44 to 1800 psi) 5 to 250 bar (73 to 3600 psi) 8 to 350 bar (116 to 5000 psi)
Internal leakage	<5 drops/min at 420 bar (6000 psi)
Fluid operating temperature	-29° to +60°C (-20° to 140°F)
Fluid Compatibility	Mineral-based or synthetic
Viscosity	7.4 to 420 cSt (50 to 2000 SUS)
Filtration	21/19/16 or better (per ISO 4406) Use with a filter rated ß10 ≥ 200
Cartridge Material	Steel with hardened work surfaces Zinc-plated exposed surfaces
Seal material	Buna N or Viton® O-rings Solid thermoplastic polyester backup rings
Threads	Metric, SAE

Chapter 12 | Advantages of Cartridge Valve Systems

The controls in today's machinery require exceptionally affordable and energy-efficient control systems. The benefits of cartridge valve systems over traditional hydraulic systems include the following: smaller system size, enhanced system response, reduced system costs, reduced chance of leakage, fewer pressure shocks, lower sensitivity to contamination, improved service life, more cost-effective control, and reduced energy consumption. These factors are elaborated in the following sections.

General
Integrated manifolds are compact, fast-acting, leak-proof, and economical.

System Size
Cartridge valves can be installed in custom-designed manifolds to create integrated manifolds. Each cartridge can be sized for actual flow. Cartridge valves can help reduce the cost and overall size of the system, which is particularly useful when space is at a premium.

System Response
Cartridge valves start the flow as soon as each cartridge poppet lifts off its seat. This means they can greatly enhance system response compared to spool-type valves, providing more precise control over the fluid flow. Cartridge valves also enable faster machine cycling, increasing productivity.

Pressure and Flow Capacity
They offer higher flow capacity, higher permissible operating pressures, lower pressure drops, increased power density, and lower pressure drop. Higher flow capacity results in more economical valve sizing.

Control
Cartridge valves help build inexpensive and flexible circuits that are compact, reliable, and leak-free. Multi-function cartridge valves can reduce the number of control components required, resulting in lower machine costs and simpler installation. A cartridge valve system in a machine can provide exceptional load control and safety features, resulting in improved machine performance. Also, each port in a cartridge valve can be controlled individually.

Pressure Shocks
Each cartridge can be tuned to reduce shocks, which can cause damage to the system and its components.

Leakage
Cartridge valves use a positive-seating poppet mechanism that minimizes internal leakage and energy wastage. An integrated manifold eliminates piping, providing leak-free construction.

Further, the cartridge valve design approach, with the complete control system in a single manifold, tends to reduce the possibility of external leakage.

Reducing leakage in the cartridge valve system also minimizes energy wastage.

Contamination
Cartridge valves are highly resistant to contamination.

Energy Consumption
Lower pressure drops in cartridge valves result in less heat generation and greater machine efficiency.

System Costs
Integrated manifold systems reduce the number of valving components, lowering machine costs. They also reduce installation and maintenance costs by eliminating the hoses, tubes, and fittings required with traditional hydraulic valves.

Installation and Maintenance
Cartridge valve systems are compact packages that simplify machine plumbing. Cartridge valves must be installed in standardized cavities that precisely accommodate the specific valve.

Hydraulic systems with surface-mounted cartridge valves can be easily serviced without disturbing system piping. This increases machine uptime and reduces maintenance time and costs.

Operation in Hazardous Environment
Cartridge valves are effective in hazardous environments.

Chapter 13 | Applications of Cartridge Valve Systems

The cartridge valve concept is the modern approach to hydraulics. Cartridge valves are essentially 2/2-way logic valves that perform various functions depending on the control input. They can be used as directional control valves for start, stop, and directional control functions. They can also be used as pressure control valves for pressure relief, pressure control, pressure sequence, and unloading functions.

Furthermore, they can be used as check valves for both the check and pilot-operated check functions. The preferred mounting mode is the manifold block, which can accommodate several valves depending on the hydraulic circuit for the specific application. The valve is assembled after machining the manifold and preparing the cartridges in accordance with the requirements.

Hydraulic Application Requirements

In today's manufacturing industries, there is a growing need for high-performance hydraulic systems to meet the demands of an increasingly competitive and environmentally conscious world. These systems must deliver higher power, greater productivity, outstanding quality parts, and reduced operating costs. They should also be designed to optimize manufacturing, reduce waste, and minimize environmental impact.

They should provide higher reliability and precision, lower energy usage, and greater safety, probably at extreme operating temperatures. They should preferably be cost-effective, modular systems.

How to Meet the Application Requirements?

When properly combined, cartridge valves can provide flexible design solutions for many hydraulic applications. They are essentially high-flow poppet or spool elements controlled by small pilot devices. They can control flow, pressure, or direction and perform multi-task control functions. Integrated cartridge valves mounted on a manifold can simplify machine design, reduce system costs, and provide a compact circuit. By combining cartridge valves with other hydraulic components such as accumulators, filters, and accessories, integrated manifolds can reduce size, cost, and complexity, helping design cost-effective hydraulic machinery and equipment.

Cartridge valves, with their unique design and flexible operation, can speed up machine cycles, enhance efficiency and productivity, minimize heat dissipation, reduce the weight of hydraulic plants, and provide easy installation and service.

Cartridge valves, including integrated manifolds, are commonly used in hydraulic systems across various industries, from plastics manufacturing to metal forming and iron and steel production, to meet specific application requirements.

Cartridge valves are the preferred option for medium-to-high-pressure industrial and mobile hydraulic systems, especially when flow exceeds 150 lpm (40 gpm), to control motion and power transmission across various applications. As hydraulic engineers become aware of the design flexibility, excellent performance, and substantial operational savings of cartridge valves, they are being used in more installations.

Applications, Cartridge Valves

Cartridge valves can be used in various ways to create a variety of hydraulic control functions suitable for many applications. The typical applications of cartridge valves and integrated manifolds include the following:

-Presses
-Steel mill machines
-Die-casting machines
-Metal forming machinery
-Injection molding machines
-Machine tools
-Mobile vehicles
-Construction Equipment
-Farm machinery
-Lift trucks
-Road maintenance equipment
-Marine systems
-Container handling
-Shovel loaders
-Forestry
-Dump trucks

Chapter 14 | Installation and Maintenance of Cartridge Valve Systems

When it comes to hydraulic systems, choosing components and systems that meet all necessary performance, maintenance, and safety requirements for an application is crucial. It is equally important to follow a standardized approach or adhere to the manufacturer's recommended guidelines to maintain the system while adhering to all necessary safety precautions. In the following sections, we outline preventive measures and maintenance steps to ensure the reliability and safety of hydraulic systems. It is worth noting that these recommendations are not an exhaustive list of safety and maintenance points. Instead, they provide a starting point for consideration. The designer must consider all local operating and environmental conditions and applicable regulations in their respective regions.

Precautions, Cartridge Valve Hydraulic Systems

Here are some general and application-specific steps to keep in mind when working with hydraulic systems that have cartridge valves:

- Ensure that only trained and competent personnel work on hydraulic systems.
- Use personal protective equipment to avoid accidental injuries during installation or maintenance.
- Pay attention to the component surface temperature during maintenance and troubleshooting.
- Do not grab or handle products from moving parts such as cables, levers, the upper side of cartridges, etc.
- Remember that all valves or groups of valves are attributable to pressure vessels, so placing the components in a closed but ventilated compartment is always recommended. This can protect users and the environment in the event of the accidental ejection of materials such as fittings, pipes, plugs, expanders, etc., under pressure.
- It is strongly recommended that all hydraulic pressure be relieved from the system before removing or disassembling components or parts such as pressure gauge ports, purge plugs, etc.
- Please remove tension from the coils before any maintenance or installation operation.

Installation, Cartridge Valve Hydraulic Systems
Here are some tips to keep in mind when installing cartridge valves:
- Always refer to the circuit diagram and identify each port by its name.
- Before assembling, flush clean all hydraulic circuit components.
- Allowing the pump to rotate backward is strictly prohibited, even when it is started for the first time.

Maintenance, Cartridge Valves
Cartridge valves are precision-manufactured components that require careful handling. They are susceptible to wear and damage, and defects in the valve's cavity can cause distortions in the body. Additionally, the valve's threads are a potential source of contamination, as metal particles may break off during tightening.

If a cartridge valve is faulty, replacing it with a new one of the same type is advisable. Strip parts of old cartridges to detect any signs of damage, as this may indicate other faults. If you notice score or rubbing marks on the sides of the spool, it could indicate issues with the bore's straightness or eccentricity.

Proper contamination control is necessary to maintain the cartridge valves in good working condition. This will ensure their long service life and reliable operation. Remember, pressure and return line filters enhance system life, reduce maintenance, and lower costs. Follow the outlined maintenance points to avoid costly repairs and replacements while ensuring optimal performance.
- Keep new cartridge valves in sealed bags until they are fitted
- Check the cavity of a new cartridge valve for dirt
- Re-wash them in an appropriate cleaning solution
- Make sure all cartridge surfaces are clean and free from contamination
- Ensure all seals and backup rings are in perfect condition and correctly placed
- Use only standard seals for sealing purposes.
- Check that the cavity is clean without sharp edges or chips
- Dip the cartridge insert in clean oil
- Screw or slip the insert into the cavity
- Tighten the insert sufficiently
- Check seals for wear when removing a cartridge valve
- Change the hydraulic fluid as and when required

15 Objective Type Questions

1. A cartridge valve system consists of:
   a) An insert, a manifold, and a control cover
   b) 4/2 - directional control valve, LVDT, and an amplifier
   c) 4/2 - directional control valve, transducer, and an amplifier
   d) 4/2 - directional control valve, torque motor, and an amplifier

2. Mark, the class of the hydraulic system, where a bank of valves is integrated into a block
   a) Servo valve system
   b) Cartridge valve system
   c) Proportional valve system
   d) Conventional electro-hydraulic system

3. The nominal bore sizes of cartridge valves are:
   a) 10, 20, 40, 50, 60, 80, 100, and 160 mm
   b) 10, 20, 30, 40, 50, 60, 80, and 160 mm
   c) 16, 25, 32, 40, 50, 63, 80, and 100 mm
   d) 20, 25, 30, 40, 50, 60, 80, and 100 mm

4. Multifunction cartridge valves help to provide:
   a) Check function only
   b) Check and directional functions
   c) Check, directional, and flow control functions
   d) Check, directional, flow control, and pressure control functions

5. Cartridge inserts in systems for pressure control functions use an area ratio of:
   a) 1:2
   b) 1:1.6
   c) 1:1.1
   d) 1:1

Note: The answer key is provided towards the end of the following 'Review Questions' section.

16 | Review Questions

1. What are the disadvantages of conventional hydraulic systems using spool-type valves?
2. Write two significant advantages of the hydraulic cartridge valve systems.
3. Explain the basic concept of a hydraulic cartridge valve with a neat sketch.
4. What are the essential parts of a cartridge valve system? Describe.
5. Differentiate the screw-in type and the slip-in type cartridge valves.
6. Distinguish between the multifunction cartridge valve and the cartridges in an integrated manifold.
7. Enumerate three important features of cartridge valves.
8. Explain the working of a single-function cartridge valve with the help of its symbolic representation.
9. What are the uses of the control cover used in a hydraulic cartridge valve?
10. Why is an orifice provided in the control cover flow path to the spring chamber of a cartridge valve?
11. What are cartridge valves' standard area ratios ($A_A:A_{AP}$)?
12. What are the design options for the poppet of a cartridge valve?
13. Draw the symbol of a cartridge valve, marking the critical areas. What is the significance of these areas?
14. What are the construction materials for the cartridge valve's body?
15. What seal materials are used to construct cartridge valves?
16. Draw the symbol of an opening cartridge valve with two working areas.
17. Draw the symbol of an opening cartridge valve with only one working area.
18. Draw the symbol of a closing cartridge valve.
19. Draw the basic directional control cover symbol with a control port X.
20. Draw the symbol of a cover with a 4-way pilot interface.
21. Draw a basic cartridge valve circuit for the port A to port B directional control of a hydraulic system using a control cover with only a pilot port X.
22. Explain how a closing cartridge valve with a basic control cover controls a hydraulic motor.
23. Write a brief note on the control covers of cartridge valves.
24. Develop a cartridge valve circuit with a cover incorporating a pilot-operated check valve for a check function when the pilot signal is absent and a directional function when the pilot signal is present.
25. Develop an electrohydraulic circuit for the ON/OFF control of a hydraulic motor using a cartridge valve, external piloting, and a 3/2-way solenoid valve. Use a proper cover with pilot interfacing.

26. Develop a cartridge valve circuit with a cover to realize a check function.
27. Develop a circuit for the ON/OFF control of a hydraulic motor using a cartridge valve employing external piloting.
28. Explain the operation of a multifunction cartridge valve with a built-in shuttle valve in its control cover. Draw its circuit representation.
29. Explain the operation of a multifunction cartridge valve with a control cover for mounting a directional control valve. Draw its circuit representation.
30. Draw a cartridge valve circuit for a hydraulic motor's ON/OFF function using a 4/2-way solenoid-operated pilot valve. The cartridge must be dumped when the pilot valve is energized.
31. A fixed displacement pump controls a hydraulic motor through a cartridge valve system. The system utilizes a control cover with a shuttle valve and a pilot interface. A 4/2-way solenoid-operated pilot valve is employed to achieve a shut-off function in the normal position and a check function (port A to port B free flow) when the valve is actuated. Design a cartridge valve circuit with a shuttle valve in the control cover.
32. Draw the circuit to realize the control function of the center position of the open-center 4-way directional control valve with cartridge valves.
33. Draw the circuit to realize the 'P to A and B to T' switching function of a 4-way directional control valve with cartridge valves.
34. How can the basic cartridge valve be modified to use it as a check valve? Explain. Draw the circuit representation of the resultant valve.
35. Develop a cartridge valve circuit for the directional control of a double-acting hydraulic cylinder using 2/2-way cartridge valves in a bridge circuit arrangement and 3/2-way solenoid-operated directional control valves for the pilot control.
36. Develop a cartridge valve circuit for the directional control of a double-acting hydraulic cylinder using 2/2-way cartridge valves in a bridge circuit arrangement and a 4/3-way solenoid-operated directional control valve for the pilot control.
37. Draw the circuit diagram of a cartridge valve system for the pressure relief and venting functions using an insert and a cover, and briefly explain.
38. Explain a basic cartridge valve system for the pressure relief function.
39. Explain a basic cartridge valve system for the unloading function.
40. Develop a hi-lo hydraulic circuit using a cartridge valve system.
41. Explain a basic cartridge valve system for the pressure-reducing function.

42. A hydraulic system using a fixed-displacement pump and a cartridge-type unloader is to be designed to store energy in an accumulator. The pump automatically unloads when the accumulator reaches the required pressure. Develop a control circuit.
43. Explain the constructional features of a cartridge for a flow control function.
44. Explain the operation of a multifunction cartridge valve with a stroke limiter. Draw its circuit representation.
45. Draw the circuit diagram of a cartridge valve system for the check function with an adjustable stroke limiter using an insert and a cover, and briefly explain.
46. Differentiate poppet-type and spool-type cartridge valves.
47. Explain a 3-way spool-type cartridge valve.
48. Draw two symbols of 3-way spool-type cartridge valves.
49. Describe a 4-way spool-type cartridge valve.
50. Draw two symbols of 4-way spool-type cartridge valves.
51. What is an actively controllable cartridge valve?
52. Explain a 2-way actively controllable Cartridge valve with a symbol.
53. What is a standard step poppet in an actively controllable cartridge valve?
54. What is a step poppet with a small seat in an actively controllable cartridge valve?
55. What is a step poppet with a seat and a dampening nose in an actively controllable cartridge valve?
56. Draw symbols of three types of poppets in actively controllable cartridge valves.
57. Draw two types of control covers for actively controllable cartridge valves.
58. Draw the symbol of an active cartridge valve with an intermediate control cover suitable for a 4-way solenoid-operated pilot valve to obtain a normally open function.
59. Draw an active cartridge valve system configuration with a cartridge, intermediate cover, control cover, and pilot valve.
60. What are the essential features of active cartridge valves?
61. Write a brief note on the applications of active cartridge valves.
62. Draw an electrohydraulic circuit to control an accumulator using an active cartridge valve system for energy storage.
63. How can proportional flow control be realized? Explain.
64. Explain three symbolic representations of the proportional flow control function.

65. Explain the proportional in-line pressure-compensated cartridge-style flow control valve.
66. Explain the electro-hydraulic proportional priority pressure-compensated cartridge-style flow control valve.
67. Describe the proportional in-line, non-compensated cartridge-type flow control valve
68. Explain the proportional 3/2-way throttle cartridge with a symbolic representation.
69. How can the proportional pressure control function be realized using a cartridge valve?
70. Draw a proportional pressure relief function configuration in the cartridge version.
71. Draw a configuration of a proportional pressure-reducing function in the cartridge version.
72. Explain the constructional features of an integrated manifold system with many cartridge valves.
73. What are the necessary steps for designing and constructing an integrated manifold system?
74. Explain the machining and assembly stages of the integrated manifold.
75. What are the advantages of integrated manifolds?
76. What are the applications of integrated manifolds?
77. What are the standard bore sizes of cartridge valves?
78. Enlist typical specifications of cartridge inserts and covers.
79. What are the advantages of the cartridge valve systems?
80. State some critical applications of the cartridge valves.
81. What are the typical modern industrial hydraulic application requirements?
82. What are the advantages of cartridge valves for meeting the challenging requirements of modern hydraulically operated machinery and equipment?
83. What are the precautions to be taken while handling hydraulic systems?
84. List four installation points about cartridge valve systems.
85. List four important maintenance aspects of cartridge valve systems.
86. The system with a fixed-displacement pump must limit the pressure in a hydraulic line when the single solenoid valve is energized and vent the cartridge when the solenoid valve is de-energized. Develop the control circuit to realize the function.

Answer key: Objective type Questions: *1-a, 2-b, 3c, 4-d, 5-d*

Appendix 1

Symbols of Cartridge Inserts

Figure A1.1 shows symbols of cartridge inserts.

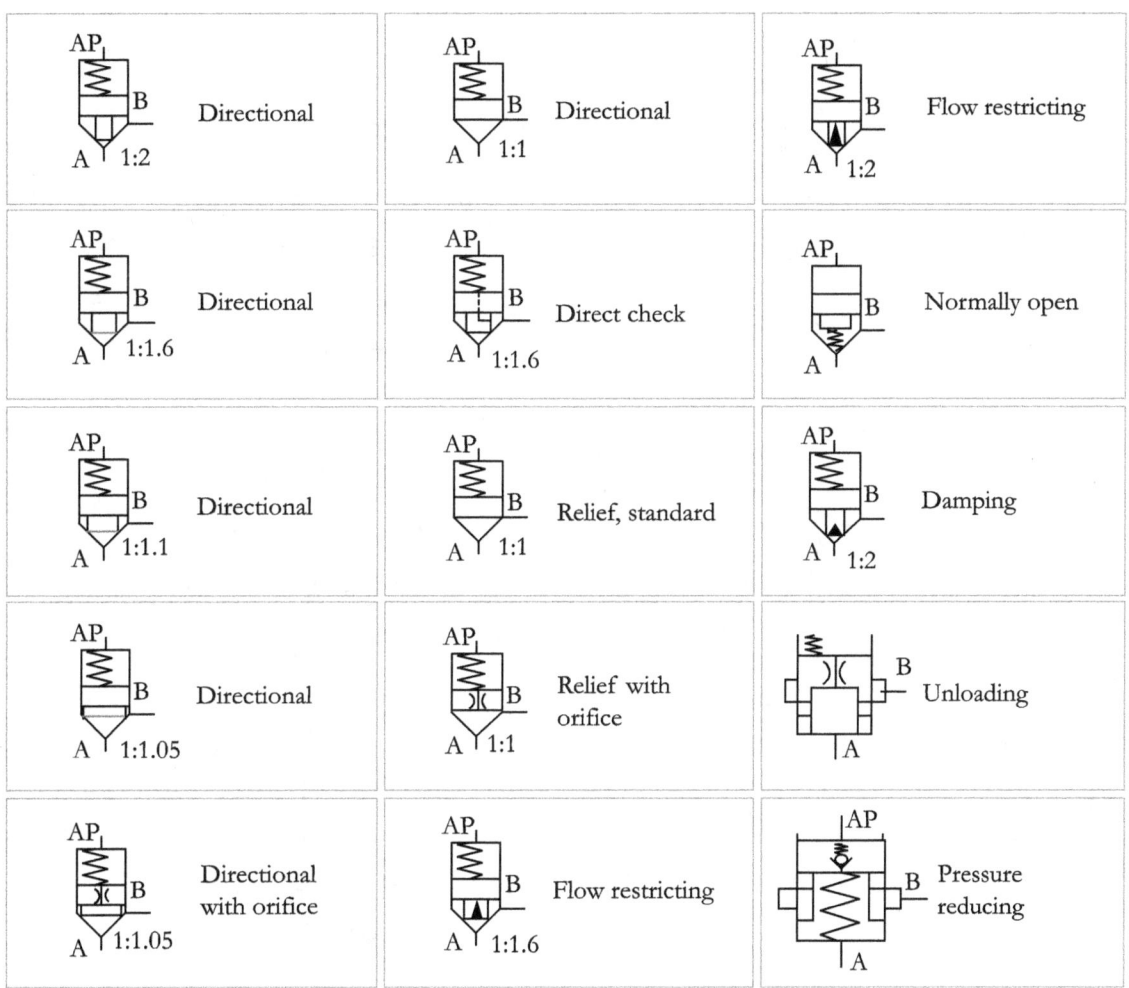

Figure A1.1 | Symbols of cartridge inserts

Note: Area ratio $A_A:A_{AP}$, if any, is indicated alongside each symbol.

Appendix 2

Symbols of Cartridge Covers

Figure A2.1 shows symbols of cartridge covers.

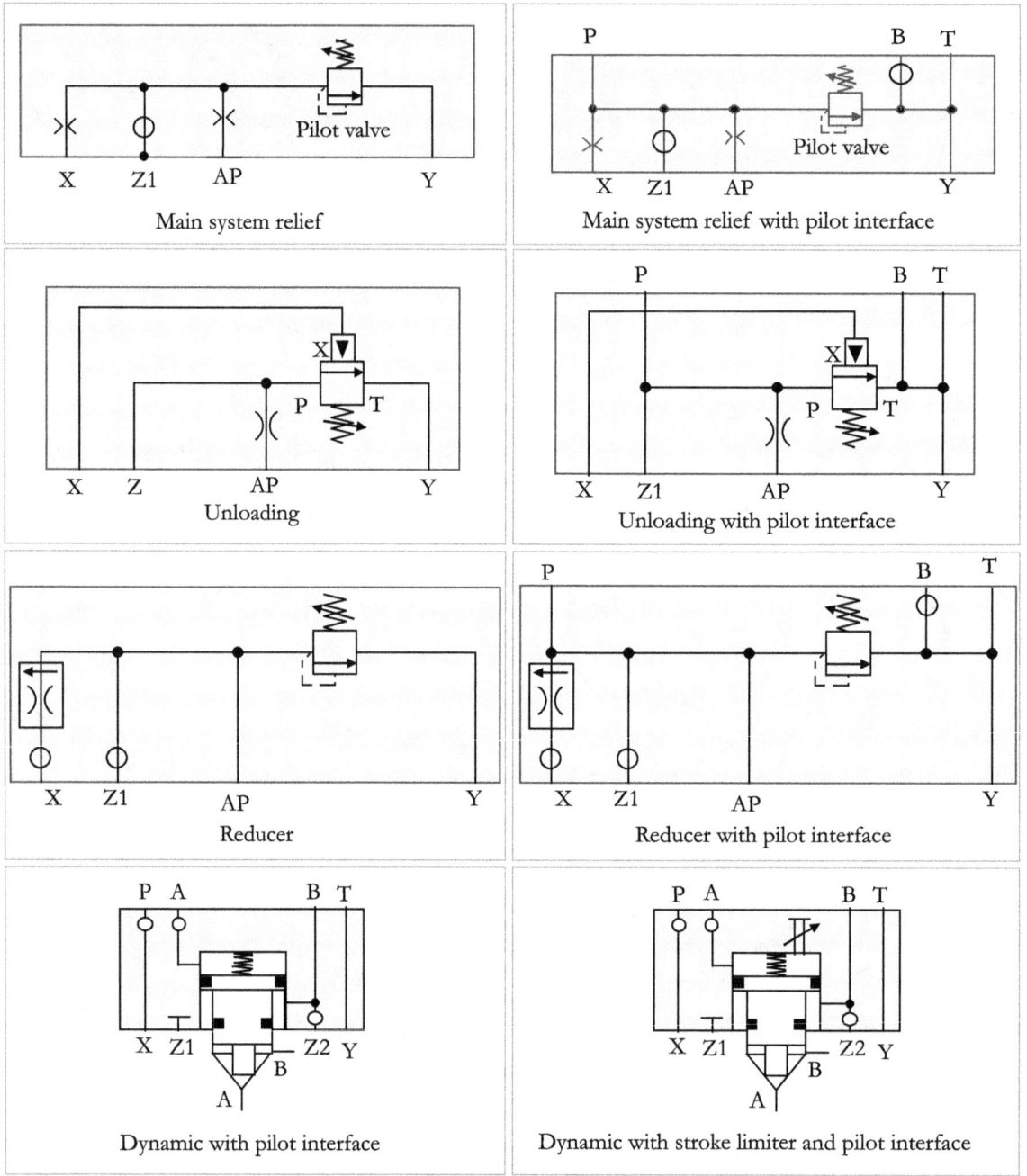

Figure A2.1 | Symbols of cartridge covers

Figure A2.2 shows symbols of cartridge covers.

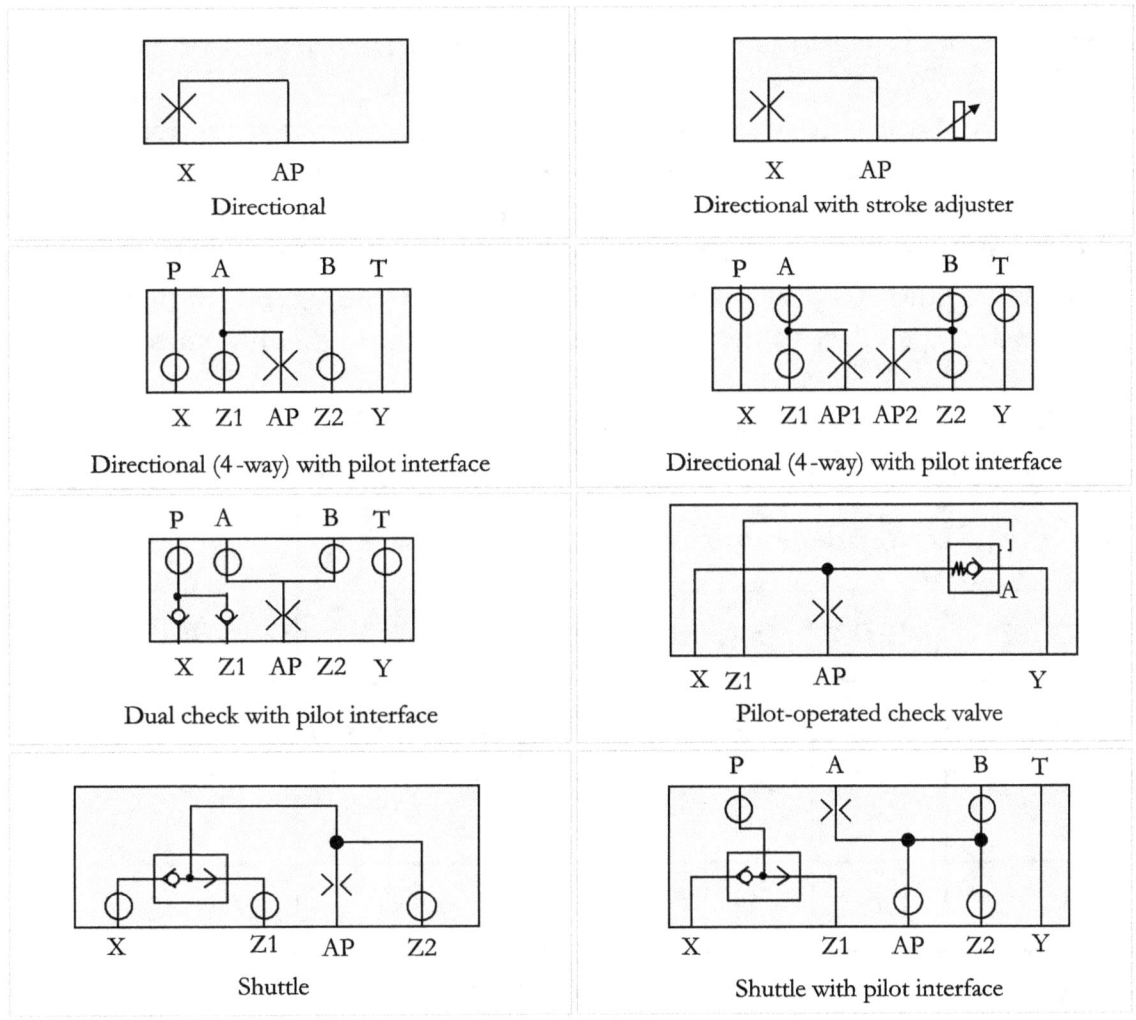

Figure A2.2 | Symbols of cartridge covers

Appendix 3

Mounting Configurations of 4-port Hydraulic Directional Control Valves

A proportional valve to be connected to a hydraulic system should be sized according to the maximum flow rate estimated for the system. Table A3.1 presents the size designations, port sizes, and nominal flow rates for different 4-port hydraulic directional control valve sizes according to the DIN (NG part), NFPA, ISO, and CETOP standards. However, the given dimensions are indicative, and the exact values can be obtained from the manufacturer's catalog.

Table A3.1 | The size designations, port sizes, and nominal flow rates for different sizes of directional control valves

Size representations				Port dia inch (mm)	Nominal flow gpm (lpm)
NG*	NFPA	ISO	CETOP		
NG 4	D02	02	2	0.177 (4.5)	5 (20)
NG 6	D03	03	3	0.295 (7.5)	10 (40)
NG 10	D05	05	5	0.44 (11)	20 (80)
NG 16	D07	07	7	0.69 (17.5)	30 (120)
NG 25	D08	08	8	0.984 (25)	60 (240)
NG 32	D10	10	10	1.25 (32)	100 (400)

A directional control valve, or a group of valves, for a hydraulic system can be configured in many ways to suit the required installation convenience. According to the way the valve body and ports are organized, the valve or valve system can be of the following types: (1) line-mounted, (2) sub-plate mounted, and (3) manifold-mounted.

Line-mounted Valves

In a line-mounted valve, the valve assembly includes the valve body and ports as an integral unit, as shown in Figure A3.1. The ports are threaded to fix fittings for fluid conductors. Therefore, the conductors can be directly connected to the valve.

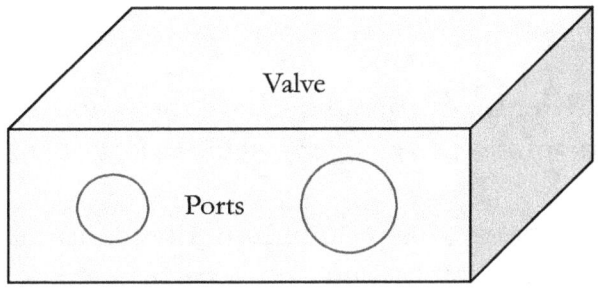

Figure A3.1 | A line-mounted valve

Line-mounted valves are lightweight and less expensive. However, they are prone to leakage. Further, it is not easy to assemble or disassemble a line-mounted valve, as all connections to the valve must be removed when it is repaired or replaced. Line-mounted valves are suitable for mobile equipment and small-flow hydraulic systems.

Sub-plate Mounted Valves

As shown in Figure A3.2, the valve and set of connection ports are distinct sections in a subplate-mounted valve. All the ports are provided on a subplate, which can be side-ported or bottom-ported. All conductor connections are made to the ports on the subplate.

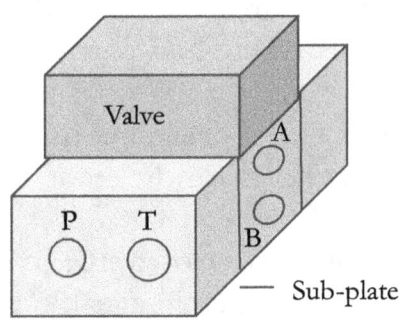

Side ports: P, A, B, T

Figure A3.2 | A sub-plate mounted valve

The subplate serves as a convenient mounting pad for one valve. It contains bores, mostly with a standard pattern, to pass a fluid medium and realize the associated valve's control function.

The valve with O-ring seals is mounted to the subplate using bolts. The seals are necessary to eliminate leaks.

Valve manufacturers offer many thread options, including NPT, SAE, metric, and BSP. Sub-plates are manufactured as per a standard or custom-made.

Depending on the system pressure, aluminum, ductile iron, or steel material can be used to construct a subplate. Aluminum can be used at pressures up to 210 bar, and ductile iron at pressures up to 350 bar.

Disconnecting conductor connections is not required when replacing a valve mounted on a subplate. This feature is convenient, as it can greatly reduce the time and cost of replacing the valve. Some manufacturers offer wiring channels in the subplate.

The subplates come in many sizes, patterns, ports, mounting hole locations, and pressure ratings.

Interface Layouts for Sub-plates

The sizes, locations, and pattern of ports and mounting holes on the mounting surface of a sub-plate should perfectly match that of the associated four-port hydraulic directional control valve. Therefore, the parameters for the mounting surfaces of valves and sub-plates are standardized in accordance with NFPA T3.5.1 MR1, ISO 4401, CETOP, or the NG part of DIN 24340.

These standards specify sizes, size designations, interface layouts, and the locations of ports and mounting holes for different valve and sub-plate sizes.

A sub-plate-mounted directional control valve conforming to a particular standard from any manufacturer is interchangeable with a valve of comparable size and conforming to the same standard from a different manufacturer.

The probable difference can be whether the bolts have SAE or metric threads.

Interface Layout for a Sub-plate of Size 02 as per ISO 4401 [NFPA D02, CETOP 2, or NG 4]

Figure A3.3 shows the locations, port patterns, and holes for mounting bolts and locating pins on the mounting surface of a sub-plate for size 02, as per ISO 4401 [NFPA D02, CETOP 2, or NG 4].

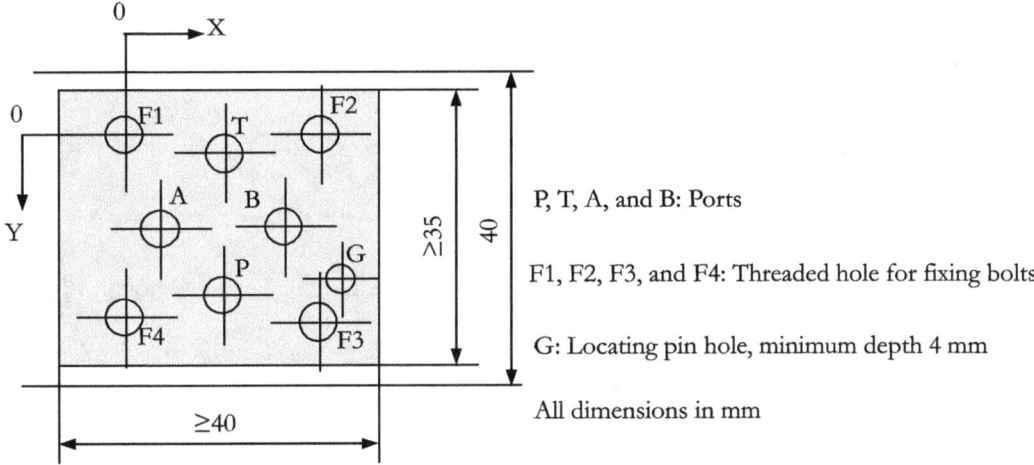

Figure A3.3 | Interface layout for a sub-plate of size 2 as per ISO 4401

Port Sizes and Locations, Size 02 Conforming to ISO 4401

Table A3.2 gives the indicative sizes and locations of ports and other openings for fixing bolts and locating pin G.

Table A3.2 | Sizes and positions of ports and holes for mounting bolts and locating pins for size 02, ISO 4401

Axis	P	A	T	B	F1	F2	F3	F4	G
	Φ 4.5 max	Φ 4.5 max	Φ 4.5 max	Φ 4.5 max	M5	M5	M5	M5	Φ 3.4
x	12	4.3	12	19.7	0	24	24	0	26.5
y	20.25	11.25	2.25	11.25	0	-0.75	23.25	22.5	17.75

Interface Layout for a Sub-plate of Size 03 as per ISO 4401 [NFPA D03, CETOP 3, or NG 6]

Figure A3.4 shows the locations, pattern of ports, and holes for mounting bolts and locating pins on the mounting surface of a sub-plate for size 03 as per ISO 4401 [NFPA D03, CETOP 3, or NG 6] without pilot ports. The details of the sub-plate for size 03 with pilot ports are given in Figure A3.5.

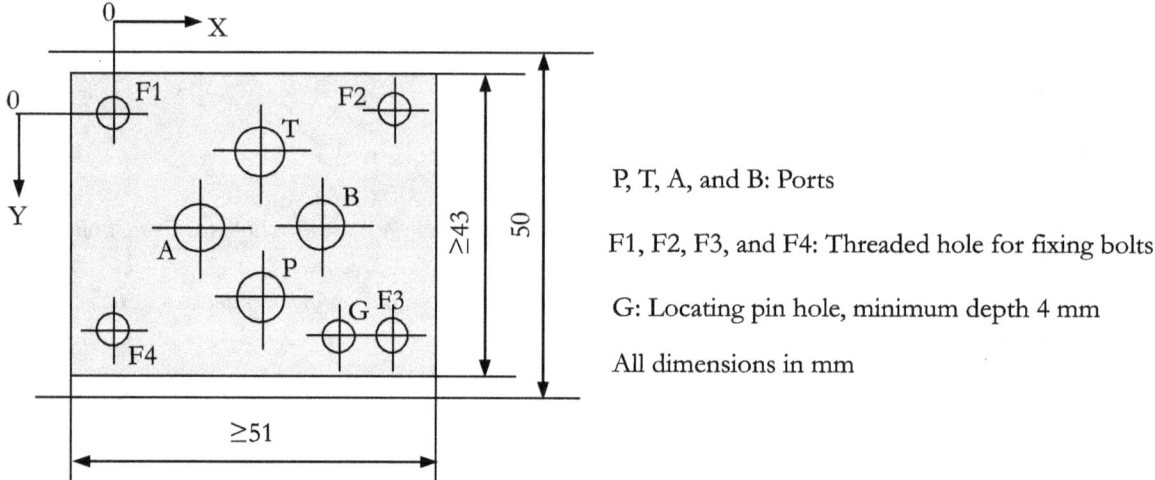

Figure A3.4 | Interface layout for a sub-plate of size 03 (without pilot ports) as per ISO 4401

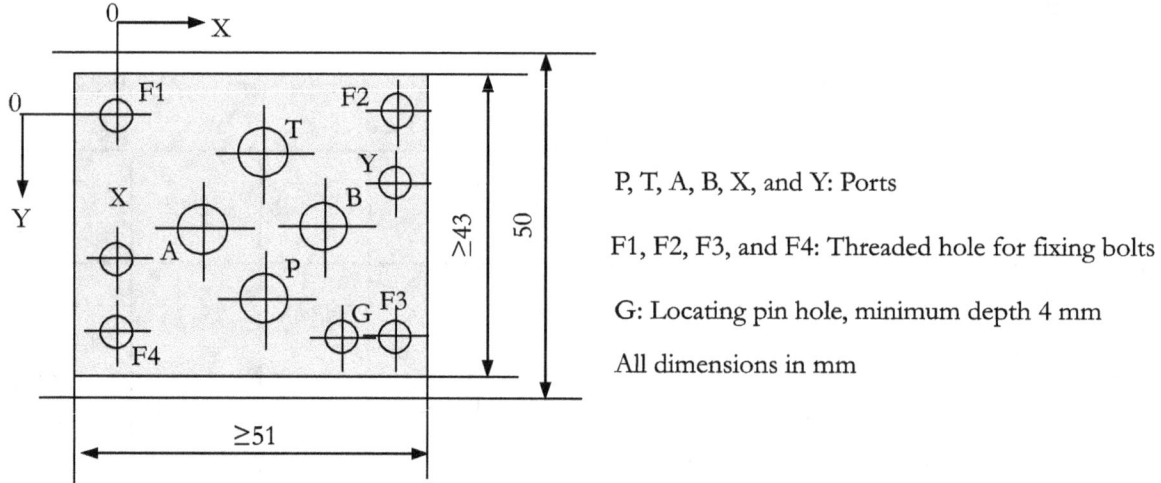

Figure A3.5 | Interface layouts for sub-plate for size 03 (with pilot ports) as per ISO 4401

Port Sizes and Locations, Size 03 Conforming to ISO 4401

Table A3.3 gives the indicative sizes and locations of ports and other openings for fixing bolts and the locating pin.

Table A3.3 | Sizes and positions of ports and holes for mounting bolts and locating pins for size 03, ISO 4401

Axis	P	A	T	B	F1	F2	F3	X	Y	G
	Φ 7.5 max	Φ 7.5 max	Φ 7.5 max	Φ 7.5 max	M5	M5	M5	Φ 3.3 max	Φ 3.3 max	Φ 4.
x	21.5	12.7	21.5	30.2	0	40.5	40.5	0	40.5	33
y	25.9	15.5	5.1	15.5	0	-0.75	31.75	22	9	31.75

Manifold Assembly

The flow in a complex hydraulic system with conventional pipe connections tends to be restricted. Further, the pipe connections can become potential leakage points.

A hydraulic system with a manifold assembly enables the creation of hydraulic circuits without the use of pipes and fittings. It helps build a compact, leak-free system that is easier to maintain.

A single-piece bar manifold or stackable plate assembly in a hydraulic system provides a single place to mount several valves with standard mounting patterns. These units also have wiring channels and plug-in valves for solenoid operation.

The manifold in a hydraulic system is designed to distribute fluid throughout the system. The flow of the pressurized fluid is regulated by hydraulic valves installed within the manifold.

As shown in Figure A3.6, the bar manifold supports all valves and contains all the passages for the entire hydraulic system.

A stackable modular plate assembly consists of two or more sub-plates connected to form a valve stack with internal passages for a common pressure connection and a common tank connection.

Each modular subplate unit supports only one valve and contains internal passages for the supported valve and flow-through provisions. It is normally connected to a series of similar modular blocks to form a complete system.

Figure A3.6 | A bar manifold

There are two ways of manufacturing manifolds:

(1) A manifold can be made from a piece of steel, aluminum, or cast iron that can be drilled to provide the required flow passages.
(2) A manifold can also be custom-made from several layers of steel sheets with appropriate passages machined or milled through them.

These sheets and solid metal end plates are stacked, and the whole stack is brazed. This laminar design allows the internal passages to be contoured and as large as possible. Therefore, a rated flow rate can be permitted in a manifold with a minimum pressure drop.

Manifold systems' advantages include reduced assembly and installation costs, lower pressure drop, fewer leak points, and easy component interchangeability.

17 | References

1. Article on 'BOOK 2, CHAPTER 4: Slip-In Cartridge Valves (part 1)' in Hydraulics & Pneumatics magazine, Jan 16, 2008 edition
2. Article on 'Cartridge Valves Combining Multiple Functions into a Single Cavity', by Damiano Roberti, Application Engineer, HydraForce Hydraulics Ltd.
3. Article on 'Cartridge valves' HCV, Auckland, New Zealand
4. Article on 'Cartridge Valves' in hydraulics & pneumatics magazine, Jan 1, 2012 edition
5. Article on 'CHAPTER 11: Slip-in Cartridge Valves (Logic Valves)', in hydraulics & pneumatics magazine, Jan 16, 2008 edition
6. Article on 'Introduction to Cartridge Valves', Moog Inc., NY, USA
7. Article on 'Slip-in Cartridge Valves, MOOG Inc., USA
8. Catalog on ' Basics for 2-way cartridge valves type L1, ISO 7368 size from 16 to 100, ATOS
9. Catalog on '2/2 CARTRIDGE VALVES LOGIC ELEMENTS ACCORDING TO ISO 7368 (DIN 24342)', ARON
10. Catalog on '2/2 Logic Cartridge Valve, Size 10' BUCHER HYDRAULICS, Frutigen
11. Catalog on '2-way cartridge valve, actively controllable', Bosch Rexroth AG, Germany
12. Catalog on '2-way cartridge valves - directional function', No. RA 21 010/06.98, Mannesmann Rexroth Corporation, USA
13. Catalog on '2-way slip-in cartridge valves Directional and Pressure function, ISO 7368, Sizes 16 to 100, MOOG
14. Catalog on 'Cartridge Valves Technical Information – Introduction', SAUER DANFOSS
15. Catalog on 'Standard Cartridge Valves – 2/2 Way Series NG16 - NG100, MOOG Industrial Controls Division, USA
16. Catalog on 'Standard Cartridge Valves – 2/2 Way Series NG16 – NG100', MOOG HYDROLUX, LUXEMBOURG
17. Catalog on 'Yuken Logic Valves' Yuken
18. Catalogs on '2-way cartridge valves, Directional functions, Cartridge valves and control covers', No. RA 21010/03.05 and '2-way cartridge valves; pressure functions', No. RA 21 050/02.03, Bosch Rexroth Corp., USA

19. Catalogs on '2-way Servo Cartridge Valve ISO 7369, Sizes 40 to 100, Sizes 125 and 160', '2-way Slip-in Active Cartridge, ISO 7368 Sizes 16 to 100, '3-way Servo Cartridge Valve, Sizes 30, 50, 63', CARTRIDGE COVERS ISO 7368 SIZES 16 TO 160', MOOG

20. Catalogs on 'HY15-3502/US Contents Flow Control Valves', 'HY15-3502/US Contents Pressure Control Valves', 'HY15-3502/US Contents Proportional Valves', 'HY15-3502/US Shuttle Valves', 'HY15-3502/US, Technical Tips, Bodies and Cavities, 'HY15-3502/US Check Valves', and 'HY15-3502/US Logic Elements', No., Parker Hannifin Corporation, USA

21. Catalog on 'ACTIVE CARTRIDGE VALVES', The Oilgear Company, Milwaukee, USA

22. Catalog on 'Cartridge valves and manifolds - Pressure, Directional, Flow Check, Load, Proportional', #02087369 / 07.12 CHY1110-138, HYDAC International, HYDAC TECHNOLOGY CORPORATION, Germany

23. Catalog on 'LOGIC POPPET VALVES AND CARTRIDGE MANIFOLD SYSTEMS', ALMO Manifold and Tool Company, MI, USA. www.almomanifold.com

24. Catalog on 'Standard Cartridge Valves – 2/2 way Series NG 16 – NG 100', MOOD Industrial Control Division, NY, USA

25. Catalog on 'Vickers® Cartridge Valves - Slip-in Cartridge Valves to ISO 7368 (DIN 24342).'

26. Catalog on Slip-in Cartridge Valves, to ISO 7368 (DIN 24342), 5043.00/EN/0499/A, VICKERS

27. REFERENCE GUIDE SCREW-IN CARTRIDGE VALVE, The Oilgear Company USA

Fluid Power Educational Series Books

1. Pneumatic Systems and Circuits -Basic Level (In the SI Units)
2. Industrial Pneumatics -Basic Level (In the English Units)
3. Pneumatic Systems and Circuits -Advanced Level
4. Electro-Pneumatics and Automation
5. Design of Pneumatic Systems (In the SI Units)
6. Design Concepts in Pneumatic Systems (In the English Units)
7. Maintenance, Troubleshooting, and Safety in Pneumatic Systems
8. Industrial Hydraulic Systems and Circuits -Basic Level (In the SI Units)
9. Industrial Hydraulics -Basic Level (In the English Units)
10. Hydraulic Fluids
11. Hydraulic Filters: Construction, Installation Locations, and Specifications
12. Hydraulic Power Packs (In the SI Units)
13. Power Packs in Hydraulic Systems (In the English Units)
14. Hydraulic Cylinders (In the SI Units)
15. Hydraulic Linear Actuators (In the English Units)
16. Hydraulic Motors (In the SI Units)
17. Hydraulic Rotary Actuators (In the English Units)
18. Hydraulic Accumulators and Circuits (In the SI Units)
19. Accumulators in Hydraulic Systems (In the English Units)
20. Hydraulic Pipes, Tubes, and Hoses (In the SI Units)
21. Pipes, Tubes, and Hoses in Hydraulic Systems (In the English Units)
22. Design of Industrial Hydraulic Systems (In the SI Units)
23. Design Concepts in Industrial Hydraulic Systems (In the English Units)
24. Maintenance, Troubleshooting, and Safety in Hydraulic Systems
25. Hydrostatic Transmissions (HSTs) (In the SI Units)
26. Concepts of Hydrostatic Transmissions (In the English Units)
27. Load Sensing Hydraulic Systems (In the SI Units)
28. Concepts of Load Sensing Hydraulic Systems (In the English Units)
29. Electro-hydraulic Proportional Valves
30. Electro-hydraulic Servo Valves
31. Cartridge Valves
32. Electro-hydraulic Systems and Relay Circuits
33. Practical Book: Pneumatics - Basic Level
34. Practical Book: Electro-pneumatics - Basic Level
35. Practical Book: Industrial Hydraulics – Basic Level
36. Programmable Logic Controllers and Programming Concepts
37. Compressed Air Dryers
38. Hydraulic Circuits – Identification of Components and Analysis

For more details, please visit **https://jojibooks.com.**

About the Author

Joji Parambath has been a trainer in Pneumatics, Hydraulics, and PLC for over 25 years. During his career, he has trained numerous professionals from various industries, faculty members, and engineering students.

He is the key trainer at Fluidsys Training Centre, Bangalore, India (https://fluidsys.org), which provides training in Pneumatics and Hydraulics. He has already written two books on these subjects. The present series of 38 books is intended to restructure and update the existing books.

The author wishes to thank all trainees for their lively interaction and many useful suggestions during the training programmes that prompted the author to write the present series of books. You may send your feedback to info@fluidsys.org.

21st March 2024

www.ingramcontent.com/pod-product-compliance
Lightning Source LLC
Chambersburg PA
CBHW062221220526
45471CB00009B/3295